363.702COM
Toward Environmental Justice: Research,
Education, and Health Policy Needs

Common Environmental Justice

Toward
Environmental
Justice

Research, Education, and Health Policy Needs

Committee on Environmental Justice

Health Sciences Policy Program

Health Sciences Section

INSTITUTE OF MEDICINE

NATIONAL ACADEMY PRESS
Washington, D.C. 1999

NATIONAL ACADEMY PRESS • 2101 Constitution Avenue, N.W. • Washington, D.C. 20418

NOTICE: The project that is the subject of this report was approved by the Governing Board of the National Research Council, whose members are drawn from the councils of the National Academy of Sciences, the National Academy of Engineering, and the Institute of Medicine. The members of the committee responsible for the report were chosen for their special competences and with regard for appropriate balance.

The Institute of Medicine was chartered in 1970 by the National Academy of Sciences to enlist distinguished members of the appropriate professions in the examination of policy matters pertaining to the health of the public. In this, the Institute acts under both the Academy's 1863 congressional charter responsibility to be an adviser to the federal government and its own initiative in identifying issues of medical care, research, and education. Dr. Kenneth I. Shine is president of the Institute of Medicine.

Support for this project was provided by the National Institutes of Health through the National Institute of Environmental Health Sciences, the National Cancer Institute, the National Institute of Dental Research, the National Institute of Neurological Disorders and Stroke, the National Institute of Allergy and Infectious Diseases, the National Institute of General Medical Sciences, the National Institute of Alcohol Abuse and Alcoholism, the National Institute for Nursing Research, the National Center for Human Genome Research, the National Heart, Lung, and Blood Institute, the Office of Research on Minority Health, the National Institute for Occupational Safety and Health, and the National Center for Environmental Health (Task Order 14 of Contract No. NO1-OD-4-2139). Additional funding was provided by the U.S. Department of Energy, the Environmental Protection Agency, and the Centers for Disease Control and Prevention through an interagency agreement with the National Institute of Environmental Health Sciences. The views presented in this report are those of the Committee on Environmental Justice and are not necessarily those of the funding organization.

International Standard Book No. 0-309-06407-4

Additional copies of this report are available for sale from the National Academy Press, 2101 Constitution Avenue, N.W., Box 285, Washington, DC 20055. Call (800) 624-6242 or (202) 334-3313 (in the Washington metropolitan area), or visit the NAP's on-line bookstore at **www.nap.edu.**

The full text of this report is available on line at **www.nap.edu**

For more information about the Institute of Medicine, visit the IOM home page at **www2. nas.edu/iom.**

Printed in the United States of America

The serpent has been a symbol of long life, healing, and knowledge among almost all cultures and religions since the beginning of recorded history. The image adopted as a logotype by the Institute of Medicine is based on a relief carving from ancient Greece, now held by the Staatliche Museen in Berlin.

Committee Liaisons

ENRIQUETA C. BOND, President, The Burroughs Wellcome Fund, Durham, North Carolina
MARK R. CULLEN, Professor of Medicine and Public Health, Yale Occupational and Environmental Medicine Program, Yale University School of Medicine

Study Staff

VALERIE P. SETLOW, Director, Division of Health Sciences Policy and Study Director (until December 1997)
EDWARD HILL III, Study Director (until November 1996)
YVETTE J. BENJAMIN, Research Associate (until December 1997)
PETER BOUXSEIN, Senior Program Officer (from September 1998)
ANDREW M. POPE, Director, Health Sciences Policy Program
MARY JAY BALL, Project Assistant (until January 1997)
GLEN SHAPIRO, Project Assistant/Research Assistant (from July 1998)

Section Staff

CHARLES H. EVANS, JR., Head, Health Sciences Section
LINDA DEPUGH, Administrative Assistant
JAMAINE TINKER, Financial Associate

REVIEWERS

This report has been reviewed in draft form by individuals chosen for their diverse perspectives and technical expertise, in accordance with procedures approved by the National Research Council's Report Review Committee. The purpose of this independent review is to provide candid and critical comments that will assist the Institute of Medicine in making the published report as sound as possible and to ensure that the report meets institutional standards for objectivity, evidence, and responsiveness to the study charge. The review comments and draft manuscript remain confidential to protect the integrity of the deliberative process. The committee wishes to thank the following individuals for their participation in the review of this report:

JOHN ALDERETE, University of Texas Health Science Center at San Antonio;
MARK CULLEN, Yale School of Medicine;
CASWELL A. EVANS, JR., National Institute of Dental Research, Bethesda, Maryland;
HOWARD KIPEN, Environmental and Occupational Health Sciences Institute, Rutgers University;
JUDITH R. LAVE, University of Pittsburgh;
FLOYD MALVEAUX, Howard University College of Medicine;
GILBERT S. OMENN, University of Michigan;
ELLEN K. SILBERGELD, University of Maryland; and
ALICE S. WHITTEMORE, Stanford University School of Medicine.

While the individuals listed above have provided constructive comments and suggestions, it must be emphasized that responsibility for the final content of this report rests entirely with the authoring committee and the Institute of Medicine.

Preface

The subject of environmental justice elicits strong emotion from many parties. Neither the serious health concerns nor the charges of biased or unfair policies that are implicit in the subject can be taken lightly. More importantly, however, the communities of concern, such as those that the committee saw firsthand, carry burdens beyond poor health. They carry the burden of frustration and feelings of helplessness and betrayal. For these reasons, environmental justice differs from most other areas of research and study. The committee therefore felt that it was important to approach our task by different methods. The clearest example of this is that the committee visited a number of communities with a variety of types of exposures and potential health effects with various political, social, and regulatory histories. These interactions allowed committee members to hear firsthand the myriad interrelated concerns and to witness the residents' feelings toward systems that the residents felt did not recognize or respond to their environmental health needs.

This study by our committee was requested because of the evidence that some sectors of U.S. society bear a disproportionate share of environmental exposure and harm: more than is borne by other sectors of society and more than can be justified by any benefits that they receive from the sources of this harm. A reasonable corollary is that the concerns of these sectors are underrepresented in environmental science. Were their concerns more central to the scientific enterprise, it would have "discovered" environmental justice much earlier than has been the case.

The committee came to several important conclusions. First, it is the committee's heart-felt belief that all communities in the United States should live in environments supportive of health and that differential exposures to environmental stressors should be eradicated. Some communities exposed to higher levels of environmental stressors include minority and economically disadvantaged populations. Because the populations of these communities are small and because they also have other complex disease risks, it is frequently difficult to separate exposures to environmental stressors from other disease or health risks. For these reasons, the committee calls upon federal, state, and local regulatory authorities to work with public health agencies to ensure that no communities within their jurisdictions suffer from disproportionate exposures to any environmental stressor.

The committee's second conclusion is that environmental justice research has constituent constraints and goals. First, environmental justice research is based on solving specific problems, and therefore, the results of the research are inherently tied to the community of concern. Second, the results may need to be translated into public policy even if they are incomplete or inconclusive. The communities of concern in the context of environmental justice typically have many social, behavioral, and economic risk factors for disease as well as complex environmental stressors. This makes the identification of the causation—or determination of the etiology—of the adverse health status typically experienced by these communities difficult. However, the committee believes that a concerted effort is needed to identify approaches that can improve the ability to define causation in this context, including the increased validation and use of biological markers, the development of enhanced biostatistical and epidemiological approaches, and the provision of appropriate funding for investigators and communities that participate in this research.

The third conclusion is that the public health, medical, and policy communities—as well as the citizens of the nation—need to be given an opportunity to understand what is and what is not known about the potential for adverse health effects resulting from exposures to environmental stressors. For this reason, the committee recommends a set of educational goals that are aimed at health professionals and policy communities and that can be extended to the general public.

Addressing the environmental justice-related issues discussed above could require a substantial reorganization and reorientation of the research enterprise. It means facing core issues in how research is funded and managed. In the short run, this could mean disrupting processes that are already quite arduous. Scientists and administrators work hard to develop and implement projects within the existing constraints of lobbying the U.S. Congress, submitting proposals, executing studies, mentoring students, weathering peer review, and so on. In the longer run, though, confronting the issues raised by environmental justice can strengthen the scientific enterprise. In an era of fiscal constraint, it can focus thinking about who are the ultimate "clients" for federal research and how their political support can be secured. It may even generate a new and vocal constituency. For projects that require observations within affected communities, a cooperative relationship might be needed for research to be conducted at all. Moreover, the science itself may benefit from consideration of the social context within which it is conducted. That exercise can provide insight into unwitting assumptions that scientists might not otherwise recognize. It can provide the impetus for the interdisciplinary collaboration that must be a part of effective solutions to complex environmental problems.

Environmental justice issues and concerns typically involve several types of agencies with different research and regulatory mandates. The committee heard repeatedly from participants at the site visits about the difficulties associated with the fact that there is no well-identified point of contact in the various agencies responsible for responding to their concerns. A single federal agency or coordinating committee with better authority and responsibility across all administrative barriers should be assigned to those communities where environmental justice is a concern. An excellent example of how this could work is the consortia of agencies that have supported this study. The National Institute of Environmental Health Sciences, in taking the lead with other federal agencies (the Centers for Disease Control and Prevention, the Environmental Protection Agency, and the U.S. Department of Energy), has provided a useful approach toward a research policy request. Such an approach could provide a clearinghouse and communication channel between federal agencies and state and local entities.

Much remains to be learned about environmental health and environmental justice. Great strides need to be taken in terms of the interrelated topics of research and education before society can ensure environmental justice in its broadest sense. Until then, environmental justice needs to become a higher priority in the fields of public health, research, education, and health policy. More importantly, these areas need to be approached systematically so that research directly affects policy to improve public health and education and that policy, together with public health, identifies needs that can be addressed by research. As increasing numbers of laypeople, health care professionals, and policymakers become aware of the issues and become cognizant of the problems, communities can be assisted in striving toward environmental justice.

James R. Gavin III
Donald R. Mattison
Cochairs

Contents

Toward

Environmental

Justice

1

Introduction and Executive Summary

Each Federal agency shall make achieving environmental justice part of its mission by identifying and addressing, as appropriate, disproportionately high and adverse human health or environmental effects of its programs, policies, and activities on minority populations and low-income populations.

President William Jefferson Clinton, 1994

As an industrialized nation, the United States produces a broad range of goods and technologies that make modern life more convenient and more efficient. The same processes that generate the nation's power, manufacture its goods, and provide its transportation, however, also produce by-products that can pollute the environment and that can be hazardous to human health. The amelioration of environmental degradation in general, as well as environmental health hazards in particular, has been a prominent national concern for at least three decades. Within the last several years, an increasingly vocal concern has also been expressed: that the broad array of environmental burdens and hazards are being borne disproportionately by lower-income communities and by racial and ethnic minorities. Efforts to address this concern have been given the label *environmental justice.*

CONCEPTS AND DEFINITIONS

As defined by the Environmental Protection Agency (EPA), *environmental justice* is

the fair treatment and meaningful involvement of all people regardless of race, ethnicity, income, national origin or educational level with respect to the development, implementation, and enforcement of environmental laws, regulations, and policies. Fair treatment means that no population, due to policy or economic disempowerment, is forced to bear a disproportionate burden of the negative human health or environmental impacts of pollution or other environmental consequences resulting from industrial, municipal, and commercial operations or the execution of federal, state, local and tribal programs and policies. (Environmental Protection Agency, 1998, p. 2)

1

Environmental justice is a concept that addresses in a cross-cutting and integrative manner the physical and social health issues related to the distribution of environmental benefits and burdens among populations, particularly in degraded and hazardous physical environments occupied by minority or disadvantaged populations.

The definition of *health* adopted by the committee is that of the Constitution of the World Health Organization (1986), which defines health as "a state of complete physical, mental, and social well-being and not merely the absence of disease or infirmity." Although the health of the individual is important, much of the attention in this report focuses on what is referred to as *communities of concern*. Communities in this sense consist of groups of individuals who live, and often work, in specific neighborhoods or regions. In this report, the phrase communities of concern refers to communities that have or that are suspected of having disproportionately high levels of exposure to environmental stressors. The committee uses the term *stressors* to describe a broad range of factors that can influence human health, such as chemicals, biologics, allergens, and traditional toxicants, but it also includes light, noise, odors, and particulate matter, among others. The populations of communities of concern may also be characterized as having limited access to health care and education, being politically disenfranchised, being of low socioeconomic status, and belonging to a racial or ethnic minority group. A focus on the health of communities involves a public health perspective, defined in a 1988 report by the Institute of Medicine's (IOM's) Committee on the Future of Public Health as "organized community efforts aimed at the prevention of disease and promotion of health" (Institute of Medicine, 1988a, p. 41).

The committee defined the *environment* to include all places where people live, work, and play. This definition highlights the often-overlooked relationship between environmental and occupational health. *Environmental health*, as defined by a previous IOM report (Institute of Medicine, 1995b), is "freedom from illness or injury related to exposure to toxic agents and other environmental conditions that are potentially detrimental to human health" (p. 15). Occupational health and safety focus on the environmental conditions in the workplace. Given that low-income and minority workers in the United States are disproportionately employed in occupations with higher levels of exposure to health hazards (Frumkin and Walker, 1997) and that work-related illnesses occur in these groups at disproportionately higher rates, the relationship between environmental or occupational health and environmental justice becomes highly pertinent. Finally, the committee adopts the definition of *environmental medicine* as "diagnosing and caring for people exposed to . . . hazards in their homes, communities, and workplaces" (Institute of Medicine, 1995a, p. 8).

THE COMMITTEE'S ORIGIN AND TASK

The origin of this report lies in a series of federal efforts and activities that were designed to identify and address various issues related to environmental justice.* The first of these efforts was in 1990 at the National Minority Health Conference: Focus on Environmental Contamination. The conference, sponsored by the Agency for Toxic Substances and Disease Registry, was the first attempt by a federal agency to bring together a group of scientists who had evaluated various aspects of environmental justice from different perspectives. Since then, focused interest and coverage of environmental justice issues have accelerated with reports from EPA, congressional hearings, and reports by the U.S. General Accounting Office. In general, these activities highlight evidence that the effects of environmental health hazards are borne disproportionately by disadvantaged communities, including those who are poor, have limited education, and are either unemployed or work under hazardous conditions. Inadequate access to health care and a greater burden of disease compound whatever adverse health effects might be associated with such hazards.

The concept and goal of environmental justice gained wider recognition in February 1994, when President William Clinton signed Executive Order 12898 entitled Federal Actions to Address Environmental Justice in Minority Populations and Low-Income Populations (see Appendix D). This executive order called for each federal agency to develop programs and strategies to ensure that poor and minority communities no longer suffer from discriminatory environmental regulations or disparate environmental health effects. The signing of the executive order coincided with the Symposium on Health Research and Needs to Ensure Environmental Justice, sponsored by the National Institute of Environmental Health Sciences in conjunction with the Agency for Toxic Substances and Disease Registry and several other federal agencies. Ultimately, the proceedings of that symposium led to a request that IOM produce a report with recommendations on the research, clinical, and educational needs required to achieve environmental justice (National Institute of Environmental Health Sciences, 1994).

The Committee's Task

In response to Executive Order 12898 and the proceedings of the 1994 symposium, the National Institute of Environmental Health Sciences—as the lead agency for a consortium of other institutes of the National Institutes of Health and agencies including the Centers for Disease Control and Prevention,

---

*It should be noted that federal efforts followed others such as those of the United Church of Christ Commission for Racial Justice (1987).

the U.S. Department of Energy, and EPA—asked IOM to conduct a study that would provide an independent assessment of three general issues:

1. the specific medical and health issues that are raised by the concept of environmental justice and that require attention, for example, medical education, clinical practice and research, medical surveillance, and public health;
2. the suitable roles of basic research and medicine in addressing these issues; and
3. the appropriate priorities for medical research that would facilitate improvements in the current situation.

More specifically, IOM was asked to address the following:

• *Characterization of the medical issues related to environmental health and justice.* This would include a multiethnic focus with an emphasis on socio-economic status, the diversity of potential environmental and occupational health hazards and conditions, and an assessment and evaluation of current health surveillance systems.
• *Cost-benefit and risk-benefit analyses of environmental health and justice issues.* For this aspect of the project, case studies or proposals for study methods could be reviewed to evaluate the different types of analyses with an accent on producing new research approaches and strategies.
• *Role of emerging scientific research.* An assessment of the status and development of biomarkers of susceptibility, exposure, and effect as they pertain to characterization of the health effects associated with environmental hazards. A target could be the development of new molecular biology-based technologies and tools. Ethical and legal issues related to clinical research would also be considered, as would epidemiologic research strategies.
• *Opportunities for collaborative approaches leading to disease prevention.* This would include some specific recommendations for implementation, including strategies for optimizing the conduct of basic and applied research. Collaborative interagency regulatory strategies would also be a focus of attention here.

Thus, in general, the committee's task was to assess the potential adverse human health effects caused by environmental hazards in communities of concern and to recommend how they should be addressed in terms of public health, biomedical research, education, and health policy perspectives. This report attempts to balance scientific opportunities with public needs and emphasizes the scientific approach in balance with the recognition that the community must play an increasingly active role in decisions about research and public health interventions. The committee felt that environmental justice issues do not lend themselves well to cost-benefit analyses and therefore approaching these issues from that perspective would not be helpful at this time. Environmental health sciences research can contribute to environmental justice most effectively by identifying hazards to human health, evaluating the adverse health effects, and developing interventions to reduce or prevent risks for all members of society. Environmental justice research bears a special relationship to the communities being studied, requiring unusual degrees of collaboration if it is to be scientifically valid as well as policy relevant and if the findings are to be effectively implemented.

The Data and Site Visits

The published literature on environmental justice and the related health effects is not abundant. Indeed, very little environmental and occupational medicine research specifically includes data for communities of concern and poor or minority workers. Adequate data are not available in most instances to examine the relationships among the environmental, racial, ethnic, and other socioeconomic determinants of adverse health outcomes. More research is needed to clarify these relationships. Still, there is a fair amount of published literature on the siting of toxic waste facilities, and workplace injuries, exposures, and fatalities are the best-documented environmental effects on health. Despite the inadequacy of the information to date, it seems clear that inequities related to environmental and occupational hazards do exist.

To explore these issues in greater depth beyond what could be learned from the literature, the committee visited a number of low-income and minority communities with known environmental problems and also heard presentations from stakeholders, citizens, and other concerned parties. During these visits, committee members participated in dialogues with the residents of communities in which known putative environmental hazards existed and environmental justice issues were at the forefront. When possible, the committee also heard from local, state, and federal officials, as well as industry representatives. In each instance, the committee met with local grassroots leaders, visited the neighborhoods of people affected by environmental concerns, and heard firsthand the myriad interrelated concerns.

Some of the communities visited were highly industrialized and located in close proximity to major urban centers (Chicago and New Orleans). Other communities were located near industrialized facilities without an urban infrastructure (Nogales, Arizona) but with similar concerns about exposure. Some were in agricultural communities in which the exposures of concern were agricultural chemicals (El Paso, Texas), and some were in regions with past major federal activities (Hanford, Washington). The committee recognizes that the issues and areas of concern of the communities visited are only samples but believes that these experiences provided insightful examples that helped to shape and frame the deliberations and, ultimately, the conclusions and recommendations.

CONCLUSIONS AND RECOMMENDATIONS

On the basis of a review of the available scientific literature and the information obtained from the various site visits, the committee concludes that there are identifiable communities of concern that experience a certain type of double jeopardy in the sense that they (1) experience higher levels of exposure to environmental stressors in terms of both frequency and magnitude and (2) are less able to deal with these hazards as a result of limited knowledge of exposures and disenfranchisement from the political process. Moreover, factors directly related to their socioeconomic status, such as poor nutrition and stress, can make people in these communities more susceptible to the adverse health effects of these environmental hazards and less able to manage them by obtaining adequate health care.

The committee perceived that the minority and low-income communities that it chose to visit have had disproportionately higher levels of exposure to environmental stressors compared with those for the general population. Furthermore, the committee found that those with impoverished social, economic, and political support were also least able to effect change and create solutions for the broad range of problems that they experience. During the site visits, the committee members were told by those whom they met of the many social, economic, and geographic barriers that can separate vulnerable and minority or low-income citizens from healthy environments and adequate health care.

As a result of its deliberations, the committee formulated four overarching recommendations related to public health research, education, and health policy. Strategies for implementing the recommendations are contained in the body of the full report. Together, these constitute a framework for further action.

Research

The committee believes that a public health approach should be the central means of dealing with the environmental health problems in disadvantaged communities. Conventional epidemiology will encounter difficulties, however, because of shortcomings in existing databases, the small populations typically involved, and cultural differences of researchers with residents of the communities of concern. Public health efforts should use new and appropriately creative methods for data collection and analysis and ensure community involvement throughout the process.

Recommendation 1. **A coordinated effort among federal, state, and local public health agencies is needed to improve the collection and coordination of environmental health information and to better link it to specific populations and communities of concern.**

Public health research related to environmental justice is a legitimate and valuable endeavor and is important to the communities of concern and to national goals for improving the health of the U.S. population. To adequately address the concerns raised by environmental justice, several gaps in the current research base must be filled. These include a better understanding of the exposures to environmental hazards and susceptibilities to disease on the part of low-income and minority populations, as well as better documentation of the links between exposure and disease. Communities must participate in the identification of problems needing research and in the design and implementation of research. To address all of these needs, the committee identified three principles to help guide biomedical, environmental health sciences, and other research related to environmental justice, and makes the following overarching recommendation.

Recommendation 2. **Public health research related to environmental justice should engender three principles: improve the science base, involve the affected populations, and communicate the findings to all stakeholders (see Box 1-1).**

Education

Among health care providers and other health care professionals, residents of communities of concern, and basic environmental health researchers, there is a lack of knowledge about the specific environmental hazards for particular populations. At present, enhanced efforts in the training of health professionals and education of the public are needed. A collaborative community response to environmental risks will help detect, limit, and prevent environmental insults and their harmful health effects. Such a community response requires that health care professionals be able to diagnose diseases with environmental etiologies and distinguish between environmental and other etiologies, that the public be able to understand these risks to the health of the community, and that governmental and industrial leaders use the strengths of the community while being

---

BOX 1-1

Three Principles for Public Health Research to Address Environmental Justice Issues

1. **Improve the science base.** More research is needed to identify and verify environmental etiologies of disease and to develop and validate improved research methods.
2. **Involve the affected populations.** Citizens from the affected population in communities of concern should be actively recruited to participate in the design and execution of research.
3. **Communicate the findings to all stakeholders.** Researchers should have open, two-way communication with communities of concern regarding the conduct and results of their research activities.

responsive to their needs. Educational programs that will more effectively link all parts of the community and that will build a coherent network to meet the public's needs should be enhanced or created. Thus, the committee's overarching recommendation for education is as follows.

Recommendation 3. **The committee recommends that environmental justice in general and specific environmental hazards in particular be the focus of educational efforts to improve the understanding of these issues among community residents and health professionals, including medical, nursing, and public health practitioners. This would include the following:**

   • **enhancing health professionals' knowledge of environmental health and justice issues,**
   • **increasing the number of health professionals specializing in environmental and occupational medicine, and**
   • **improving the awareness and understanding of these issues by the general public.**

Health Policy

Good risk management decisions are based on a careful analysis of the weight of scientific evidence that supports conclusions about a problem's potential risks to human health and the environment. However, decisionmakers must balance the value of obtaining additional information against the need for making a decision with imperfect or incomplete information. Acting prematurely may needlessly end an activity or close a facility that is doing no real harm (and may be improving the local economy through taxes and jobs); it may also needlessly stigmatize a community as contaminated, which may discourage alternative future development. On the other hand, acting too slowly may expose citizens to irreversible damage from actual risks whose existence might never be proven, given the limits of scientific methods. The relative importance of the risks of acting too soon or waiting too long is a political determination. The task of researchers is to provide the best possible estimates of the long-term effects of different policies. To that end, environmental justice needs better, more policy-relevant, and better-understood science.

The committee concludes that concerns about environmental health and justice are legitimate and should be taken seriously, even if the means of addressing these concerns are still lacking in some respects the rigorous science base to which policymakers might normally aspire. Policymakers cannot assume that these concerns are without merit. However, policymakers should also recognize that many other considerations also go into such decisions as choosing a site for a new manufacturing plant or solid-waste facility, removing an alleged hazard, or imposing expensive environmental controls. Choices and trade-offs will almost always need to be made when making any decision of significant consequence. Given the current state of knowledge, the committee believes that policymakers should be attentive to potential environmental hazards and adverse health outcomes and should be meticulous about including the affected communities in the decisionmaking process.

Recommendation 4. **In instances in which the science is incomplete with respect to environmental health and justice issues, the committee urges policymakers to exercise caution on behalf of the affected communities, particularly those that have the least access to medical, political, and economic resources, taking reasonable precautions to safeguard against or minimize adverse health outcomes.**

Much remains to be learned about environmental health and justice. Great strides need to be made in terms of data collection, research, and education before society can ensure environmental justice in its broadest sense. Until then, environmental justice needs to become a higher priority in the fields of public health, research, education, and health policy. As increasing numbers of laypeople, health professionals, and policymakers become aware of the issues and become cognizant of the problems, communities will be assisted in striving toward environmental justice.

ORGANIZATION OF THE REPORT

The remainder of this report describes what the committee discovered as a result of its site visits and other data collection activities. Chapter 2 describes the evidence for the contention that exposures to environmental hazards and subsequent adverse health outcomes are borne disproportionately by communities of concern. Chapter 3 (Research) describes the application of the public health perspective to environmental justice and defines the integral roles that public health efforts and the public health community play in addressing environmental hazards. The chapter also addresses research questions and their context in environmental justice, concluding with recommendations for research and research methods. Chapter 4 (Education) focuses on the need for education and strategies that can be used to educate various sectors of the population regarding environmental risks, the prevention of disease, and the care and treatment of those exposed. It concludes with recommendations directed to the public at large and the public education system as well as to physicians and other heath professionals. Chapter 5 (Health Policy) addresses the limits of science and the effects of those limitations on policy decisions. It concludes with a discussion of the need to make policy decisions without conclusive data.

2

Establishing a Baseline

The premise of environmental justice is that communities with high concentrations of racial or ethnic minorities or low-income families are disproportionately exposed to a variety of environmental burdens and hazards. Of particular interest for this report is the specific claim that such exposures produce adverse health outcomes that are also borne disproportionately by these populations. An assessment of baseline data is therefore essential to ascertaining the relative role of environmental exposure in determining the health of a population.

This chapter first describes some of the data supporting a general conclusion that there are significant disparities in the health status of the communities of concern. It then examines what is known about disparities in exposures to environmental health hazards. Although the evidence is not as strong or as direct as that establishing disparities in health status, it is sufficient to conclude that in many communities such disparities are real and possibly quite substantial. The chapter then reviews the link between exposures to environmental health hazards and health status, with particular attention to the validity and strength of the inference that disparities in exposure are responsible, at least in part, for the disparities in health status.

DISPARITIES IN HEALTH STATUS IN THE UNITED STATES

An extensive body of literature documents the fact that not all segments of the U.S. population have experienced the same advances in health status and gains in life expectancy. Racial and ethnic minority groups, individuals of low socioeconomic status, and medically underserved populations, among others, face lower life expectancies and greater health problems than the middle- and upper-class U.S. white population (Council of Economic Advisors, 1998; National Center for Health Statistics, 1998a,b). Some of the disparities in health

11

status are associated with socioeconomic status. For example, at age 45, individuals with family incomes above $25,000 can expect to live from 3 to 7 years longer than those with family incomes below $10,000 (National Center for Health Statistics, 1998b). Also, the death rate from chronic disease for individuals ages 23 to 64 with less than 12 years of education is more than twice as high as that for comparable individuals with more than 12 years of education (National Center for Health Statistics, 1998b). Nonelderly adults living in poverty-stricken areas experience a significantly higher risk of mortality from all causes (Waitzman and Smith, 1998).

Although the health status of all U.S. racial and ethnic groups has improved steadily, disparities in major health indicators between white and non-white groups are growing. In general, African Americans, American Indians, and Hispanics are dramatically disadvantaged relative to whites in terms of most health indices, whereas Asian Americans appear to be as healthy, if not healthier, than whites in terms of some indices. These overall group differences, however, mask important differences in the health statuses of subgroups. Americans of Southeast Asian descent, for example, suffer from among the highest rates of cervical and stomach cancers of all U.S. population groups and experience poorer health overall than U.S. whites (Miller et al., 1996).

Socioeconomic status appears to operate in complex ways with race and ethnicity to account for the observed differences in health status. In general, white Americans enjoy higher incomes and education attainment levels than any other U.S. racial and ethnic group and therefore are more likely to have health insurance and to be better educated with regard to healthy behaviors and diets, are more likely to seek routine medical care, and are more likely to have better access to preventive medical services.

Socioeconomic factors, however, do not completely account for racial and ethnic differences in health status (Williams et al., 1994). Several studies indicate that racial disparities in health status persist even when controlling for socioeconomic status (Advisory Board to the President's Initiative on Race, 1998). Mounting evidence indicates that in addition to resource inequities, other factors, including discrimination in the health care system, racism-related stresses, migration, and differences in levels of acculturation may also lead to poor health among members of racial minority groups (Council of Economic Advisors, 1998; Advisory Board to the President's Initiative on Race, 1998; Williams et al., 1994). As discussed later in this report, disparities in exposure to environmental hazards are also suspected as a factor in the relatively poorer health of individuals in minority and lower-income communities in the United States.

The following examples show that significant disparities exist between U.S. racial and ethnic groups in terms of several key health indicators, even when socioeconomic differences are taken into account.

Low Birth Weight

During the period from 1989 to 1996, among women with 13 or more years of education, African American women were twice as likely as white women to give birth to low-birth-weight infants (infants weighing less than 2,500 grams [5.5 pounds]) (11.9 versus 5.5 percent). This occurred even though African American women are less likely than white women to smoke during pregnancy, which is considered an important factor in causing low birth weight. Similarly, the percentage of low-birth-weight infants was higher among American Indian women (6.0 percent), Asian or Pacific Islander women (6.8 percent), and Hispanic women (6.0 percent) than white women with similar levels of educational attainment (National Center for Health Statistics, 1998b).

Infant Mortality

From 1989 to 1991 African American women experienced an infant mortality rate over two and a half times higher than that experienced by white women with the same levels of education (13.7 versus 5.1 per 1,000 births). The infant mortality rates among American Indian women and Hispanic women with similar levels of education were 8.1 and 5.8 per 1,000 births, respectively (National Center for Health Statistics, 1998b).

Death Rates

African Americans experience higher mortality rates than whites even in areas with equivalent levels of urbanization. In large, core metropolitan areas, the mortality rate among African Americans between 1993 and 1995 was 810.5 deaths per 100,000 population, compared with a rate of 491.9 per 100,000 among the white population. This disparity in mortality is also pronounced in other geographic areas: the mortality rates among African Americans in rural and urban, nonmetropolitan areas were 737.1 and 761.9 per 100,000 population, respectively, compared with rates of 503.9 and 499.4 per 100,000 population, respectively, among whites (National Center for Health Statistics, 1998b).

Cancer

In general, African American males experience cancer approximately 15 percent more frequently than white males, with incidence rates of 560 and 469 per 100,000 population, respectively. The pattern of cancer incidence rates among males in other racial and ethnic groups is more varied, but disparities are exhibited in specific cancer sites. For example, colon and rectal cancers are more common among Alaskan Native men (79.7 per 100,000 population) and

Japanese American men (64.1 per 100,000 population) than white men (57.6 per 100,000). Racial and ethnic minorities also experience higher rates of mortality from cancer than whites (Parker et al., 1998).

DISPARITIES IN EXPOSURE TO ENVIRONMENTAL HAZARDS

Many communities contain environmental hazards that represent potential sources of health risks (for examples, see Table 2-1). Although these can affect all racial, ethnic, and socioeconomic groups, there is evidence that minorities and lower-income groups face higher levels of exposure to these hazards and, therefore, potentially higher rates of adverse health outcomes. It has been shown, for example, that non-whites are disproportionately exposed to ambient air pollutants associated with respiratory symptoms and exacerbation of other ailments (see Table 2-2).

One method of determining the potential for increased exposure is to examine the proximity of communities of concern to waste or industrial facilities. Another, more accurate way is to characterize the nature and level of exposures by either direct measurement or estimation. Both of these methods are described below.

Examining Proximity to Environmental Health Hazards

Numerous studies have shown that race is associated with increased levels of exposure to environmental hazards. The Commission for Racial Justice of the United Church of Christ published a report in 1987 entitled *Toxic Wastes and Race in the United States* (United Church of Christ Commission for Racial Justice, 1987) and updated it in 1994 (Goldman and Fitton, 1994). The major finding of these studies was that communities that had one or more commercial hazardous-waste facilities had significantly higher proportions of racial minority

TABLE 2-1 Examples of Potential Sources of Environmental Health Hazards

Sources	Substances
Agricultural runoff	Allergens
Incinerators	Heavy metals
Industrial facilities	Paints and oil wastes
Landfills	Particulate matter
Toxic-waste sites	Pesticides and herbicides
Waste treatment facilities	Radioactive wastes
	Solvents
	Volatile organic compounds

TABLE 2-2 Percentage of White, African American, and Hispanic Populations Living in Air Quality Nonattainment Areas, 1992

Pollutant	Percentage		
	White	African American	Hispanic
Particulates	14.7	16.5	34.0
CO	33.6	46.0	57.1
Ozone	52.5	62.2	71.2
SO_2	7.0	12.1	5.7
Lead	6.0	9.2	18.5

NOTE: Nonattainment areas refer to those areas that do not meet the National Ambient Air Quality Standards for various pollutants.

SOURCE: Wernette and Nieves, 1993.

populations than communities with no commercial hazardous-waste facilities. In the United States in 1993, for example, the percentage of people of color (defined as everyone except non-Hispanic whites) was 14.4 percent in zip code areas with no commercial hazardous-waste facilities, 29.5 percent in areas with one facility, and 45.6 percent in areas with three or more facilities, an incinerator, or a large landfill. A similar trend had been evident in 1980. Communities with lower per capita incomes were also more likely to be situated near commercial hazardous-waste sites as well. The original study was one of the important factors in motivating a substantial response to the environmental justice issue from the federal government.

Another national study of hazardous-waste sites used census tracts as the unit of analysis as opposed to zip codes (Anderton et al., 1994). Anderton and colleagues examined the differences in race, class, and economic indicators between the populations in census tracts with treatment, storage, and disposal facilities (TSDFs) and those without TSDFs. Although they found no difference in the mean percentage of the population that was African American, that had incomes below the poverty level, or that was on welfare, they did find differences in the percentages of the population that were Hispanic (9.4 percent in census tracts with one or more TSDFs and 7.74 percent in those with none) and that were employed in manufacturing (38.6 and 30.6 percent, respectively). Additionally, Anderton and colleagues (1994) found that the differences became more noticeable in the areas surrounding the census tracts with TSDFs—that is, tracts with TSDFs and "other tracts that have at least 50 percent of their area within a 4-kilometer (2.5-mile) radius of the center of a tract containing a TSDF" (Brown, 1995, p. 18). Using that level of analysis, Anderton and colleagues found that populations of the areas surrounding TSDFs have higher mean percentages of African Americans (24.7 percent, compared with 13.6 percent in census tracts outside the larger unit of analysis), of Hispanics, (10.7 and

7.3 percent, respectively), of people with incomes below the poverty level (19.0 and 13.1 percent, respectively), and of people on welfare (13.3 and 8.3 percent, respectively). These results are congruent with those from the United Church of Christ Commission for Racial Justice.

American Indians have also been the focus of concerns about exposure to environmental hazards. For example, uranium ore was discarded on Navajo lands in New Mexico during the 1950s (Johnson and Coulberson, 1993). An investigation by the Agency for Toxic Substances and Disease Registry found radiation levels in private residences high enough to cause health concerns, and the discarded ore was removed by the Environmental Protection Agency (EPA) and the Navajo Superfund office. Although it is not currently possible to know the extent to which the environmental concerns faced by American Indians are unique to them, additional site-specific health surveillance studies of this population are under way (Johnson and Coulberson, 1993).

The establisment of causal relationships between a health condition and the siting of environmental hazards in proximity to low-income or minority communities is a complicated and debatable exercise. Were waste sites purposely located in these communities because of discriminatory motivations, because of the lack of politically effective opposition, because land was cheap, or because of a combination of these and other factors? Were the communities characterized by the same socioeconomic and racial or ethnic indicators when the waste sites were originally established, or did the composition of the communities evolve later, as a result of economic or other factors? The economics of land values, job opportunities, and transportation undoubtedly assert a strong influence on these outcomes, and the circumstances undoubtedly vary greatly from locale to locale. For the purposes of this report, however, the committee did not believe that it was essential to try to reach conclusions about causality or motivation; no matter how a particular condition came to be, if it represents an environmental health hazard and if the burdens of such a hazard are borne inequitably, then it is appropriate to assess the scope and severity of the health burden and to evaluate potential means of ameliorating it.

Characterizing Exposure

Proximity to a source is an inexact surrogate of actual contact with toxicants from the source. Quantification of the actual emissions from the source moves the analysis a step closer to measuring actual human exposure to an environmental health hazard. This requires estimating the rate of release and the path of the material into and through the environment. For example, in one community that the committee visited (Tucson, Arizona), a common industrial practice, begun in the 1940s, was dumping organic-solvent waste into earthen ponds. This source released toxicants in two ways: (1) the solvents evaporated from the ponds, creating an airborne means of exposure for workers and the residents in surrounding communities, and (2) the solvents migrated from the ponds into the ground, where they eventually entered the subsurface water resources that the community used for its domestic water. Measurements of the concentrations in the air and water can help public health practitioners determine the total possible exposure.

The characterization of exposure in the community requires an understanding of all of the potential pathways by which pollutant releases may result in exposure. This includes direct pathways such as through the air and drinking water and less obvious pathways such as through uptake by food sources. Studies of this nature need to be sensitive to the ways in which differences in behavior, employment, and lifestyles among subgroups in the population may result in differences in exposure. For example, among the Alutiiq, Yup'ik, and Inupiat Alaskan Native peoples, the yearly intake of wild foods per person is between 171 and 272 kilograms (375 and 600 pounds). Increasing evidence of certain contaminants such as mercury in the wild food supply of these Alaskan Natives has been exhibited by methyl mercury levels that exceed those provisionally established as safe by the World Health Organization. Research that did not account for the variable ingestion of contaminated foodstuffs would fail in determining the relative effects of exposures across populations.

As the focus of analysis moves from sources and releases to environmental concentrations, the quality and quantity of the data currently available decrease significantly. Of the more than 60,000 chemicals in commercial use in the United States, data on the concentrations of chemicals in the environment are available for only a small number (National Research Council, 1984; Roe et al., 1997). One systemic gap is measures of indoor concentrations, even though the U.S. population spends much of its time indoors (Environmental Protection Agency, 1993); most monitoring systems focus on the concentrations outdoors or in the workplace.

A more accurate description of the amount of toxicants absorbed requires data on actual doses received by humans. Analysis of biological samples (e.g., urine or blood samples) from affected populations can lead to more accurate descriptions of exposure to an environmental health hazard (National Research Council, 1989b), but the samples can be difficult to obtain. Methods for tissue analysis are also limited.

The environmental toxicant for which the largest amount of data is available is lead (National Research Council, 1993). It causes a variety of health problems, including neurodevelopmental effects in infants and children and cardiovascular effects in adults. Exposure to lead can be determined by measuring the level of lead in blood. Harmful health effects have been identified for levels of lead in the blood that are greater than or equal to 10 μg/dl. There are multiple sources of lead exposure, with lead-based paint being the most common. Consequently, exposure is high in urban areas, where housing is often old and likely to be coated with lead-based paint. Blood lead levels are consistently higher for poor and minority children and for residents of central cities (Brody et al., 1994; Centers for Disease Control and Prevention, 1991, 1997).

EXAMINING THE LINK BETWEEN EXPOSURE AND HEALTH

Having concluded that there are clearly disparities in health status and that there are significant disparities in exposure to environmental hazards, the next step in analyzing the issue of environmental justice is to determine whether there are causal relationships between exposure to these hazards and health outcomes and whether the disparities in health status can be attributed to the disparities in exposure. For many of the identified potential environmental hazards, the lack of published research makes it difficult to draw a strong conclusion that disparate exposures result in disparate health outcomes.

The difficulties encountered in this type of research are illustrated in the work of Baden and colleagues (1996), who tried to examine both the interactions of multiple environmental hazards and the possible connections of these hazards to adverse health outcomes. They included in their analysis the siting of several possible environmental hazards in the Chicago metropolitan area: the large areas of waste regulated under the Comprehensive Environmental Response, Compensation, and Liability Act of 1980 (the Superfund law), waste-generating sites regulated by the Resource Conservation and Recovery Act, older existing waste disposal sites such as landfills, and emissions from highway exhaust. They conducted regression analyses of census-tract data examining the relationships between proximity to waste disposal sites and poor infant health. The research was unable to account for "causal links between hazardous-waste and ill health effects, . . . occupational exposure to toxins, parental bioaccumulation of toxins, [or] medically risky behavior such as drinking and use of illegal drugs while pregnant" (Baden et al., 1996, p. 25). The researchers also noted that "use of the census tract for analysis may introduce some aggregation bias" (p. 25). Within these constraints, they concluded that "there was no relationship between the presence of waste sites and low birth weight for children born in Chicago in 1990. . . . Neither was there one single type of waste site associated with adverse health outcomes, nor did a combination of all of the waste sites (insignificant in themselves) combine to produce adverse infant health outcomes" (p. 26). The researchers' final conclusion was, "Future research must address these issues" (p. 25). Ecologic studies such as these are subject to a variety of potential biases in addition to the aggregation bias noted by the authors, and are not by themselves adequate for making or excluding causal connections (Greenland and Robins, 1994). In spite of the general lack of published research linking disparate exposures to disparate health outcomes, some well documented links do exist. One such study used field epidemiology methods combined with a prevention intervention trial to document definitively a link between disparate exposure to dimethylformamide and disparate prevalence of toxic liver disease (Friedman-Jiménez and Claudio, 1998; Redlich et al., 1988).

Lack of Health Status Characterization

One of the shortcomings in the available literature is that most of the reported studies of environmental justice have characterized disparities only in terms of proximity to potential sources of exposure or, in a few cases, to measured exposures and have not taken the next step of trying to characterize or quantify either exposure or potential differences in health status or disease incidence between these populations and the general population (Bullard, 1990; Glickman et al., 1995; Greenburg, 1993; United Church of Christ Commission for Racial Justice, 1987; U.S. General Accounting Office, 1983; Zimmerman, 1993). Few studies have looked directly at whether local, residential exposure to environmental agents is associated with an increased incidence of disease (Sexton et al., 1993; Wagener et al., 1993).

Part of this shortcoming may be due to the fact that, except for some hazards such as lead (see Characterizing Exposure above), little is known about the physiologic or biological mechanisms by which the health hazards cause disease. Although the toxicologies of ozone and other air pollutants have been extensively studied, the mechanisms by which such gases and particles injure the lungs are diverse and not yet fully understood. The goal of fully understanding the effects of any air pollutant has been described as a "daunting task" (Brooks et al., 1995).

A large proportion of what is known about disease processes has come from research in the occupational health field. Examples include lung diseases due to dusts (e.g., silica, coal, cotton), the systemic and lung toxicities of metals (e.g., lead, mercury, cadmium, beryllium, and hard metals) and many carcinogens (e.g., asbestos, hexavalent chromates, nickel, vinyl chloride, polycyclic aromatic hydrocarbons, radon, and bischloromethyl ether), and the neurotoxicities of a variety of pesticides and solvents.

The lack of knowledge about the specific disease process does not, however, preclude the use of epidemiologic methods to explore possible associations between hazards and disease. The air pollutants listed in Table 2-2 have long been associated with clinically significant adverse health effects, including decreased respiratory function, respiratory infections, the exacerbation of asthma, chronic obstructive pulmonary disease, congestive heart failure, and increased mortality (Brooks et al, 1995). A recent National Reserch Council report (1998) describes how epidemiologic research has established consistent associations between exposure to outdoor concentrations of small particulate matter (particles smaller than 2.5 micrometers [0.006 inches] in diameter) and the adverse health effects described above. Although the biological basis for such associations is largely unknown and there is limited scientific information about the specific types of particles that cause these health effects, the results of these epidemiologic studies have been relied upon by EPA in setting national ambient air quality standards (National Research Council, 1998). Thus, at least for some air pollutants and lead poisoning, there is strong evidence to support the connection between disproportionate exposure and disproportionate health outcome.

Failure of Differentiation of Populations and Differences in Outcomes

A second shortcoming in the literature is the converse of the first, namely, that studies of adverse health effects fail to differentiate the population by race, ethnicity, or socioeconomic class and examine differences in outcomes among them (Frumkin and Walker, 1997). As noted above, occupational health studies are a primary source of knowledge about the health effects of various hazards. However, only a small fraction of the existing occupational health research has provided meaningful information on either the prevalence of occupational diseases in specific racial, ethnic, or socioeconomic population groups or the relation of race, ethnicity, and socioeconomic factors to occupational health. For example, in one systematic review of 116 studies of the epidemiology of occupational cancer published in four journals, only 14 studies (12 percent) provided data on a "non-white" group. The investigators concluded that the published literature contributes little to understanding the complex relationships among occupation, cancer, and race (Kipen et al., 1991).

One rationale given for the exclusion of minority subgroups is that the number of subjects would be too small to provide an acceptable statistical power to test the primary study hypotheses with these subgroups. (Appendix A discusses these and other related issues in greater depth.) Other reasons include difficulties in long-term follow-up; inadequate measurement, classification, and reporting of data on race, ethnicity, and relevant socioeconomic variables (Feinleib, 1993; Montgomery and Carter-Pokras, 1993); and difficulties in researcher access to high-risk workplaces that employ low-wage workers.

A Case Study: Urban Asthma

Many of the challenges posed by an analysis of environmental justice issues are illustrated by the example of urban asthma. A doubling in the rate of asthma in the United States since 1980 and its apparent association with industrialized, urbanized areas has led some investigators to suspect increased levels of exposure to environmental factors as one possible cause (Vogel, 1997). As noted above, Hispanics and African Americans are more likely than whites to live in areas where the levels of particulates, sulfur dioxide, and ozone exceed National Ambient Air Quality Standards—circumstances that may contribute to the prevalence and severity of asthma.

The prevalence of asthma appears to be more strongly correlated with lower socioeconomic status than with race and ethnicity (Institute of Medicine, 1993). However, the use of data on numbers of hospitalizations, or emergency room visits or even the rate of morbidity due to asthma as a measure of the relative impact of environmental hazards on those with low socioeconomic status is problematic, because these indicators may be strongly influenced by other factors such as lower rates of health insurance or lower levels of access to high-quality primary health care.

The rates of hospitalization and mortality due to asthma are higher in urban, low-income, or minority communities (Weiss and Wagener, 1990; Weitzman et al., 1990). For example, in New York City, African American, Hispanic, and low-income populations were found to have hospitalization and mortality rates from asthma three to five times higher than those for the general New York City population (Carr et al., 1992). African American children are three times more likely than white children to be hospitalized for asthma and asthma-related conditions and four to six times more likely to die from asthma (Mannino et al., 1998; Stapleton, 1998).

These data might be viewed as demonstrating that urban asthma is an environmental justice problem. The situation is quite complicated, however, and a conclusive finding of an environmental justice problem is not free of doubt. First, the clinical diagnosis of asthma is not entirely clear-cut, and diagnostic methods and accuracy may vary among different population groups (Gergen, 1996). Second, the etiology of asthma is complex and multidimensional. A variety of toxic, allergenic, dietary, and infectious agents, as well as genetic and acquired susceptibility factors, contribute to the disease. Environmental agents that may induce new-onset asthma or that may aggravate preexisting asthma may come from outdoor or indoor sources at home, work, school, or other locations. Studies have shown that important sources of allergens for asthmatics are household dust mites and cockroaches, which are more likely to be present in urban settings and which are more likely to be encountered by children who spend a great deal of time indoors (Vogel, 1997). Consequently, the attribution of causality to a specific environmental hazard and apportionment of causality among various potential causes are exceedingly difficult if not impossible at this time.

CONCLUSIONS

Not all segments of the U.S. population have benefited to the same extent from advances in health status and gains in life expectancy. Racial and ethnic minority groups, individuals of low socioeconomic status, and medically underserved populations, among others, face lower life expectancies and greater health problems than the U.S. white population (Council of Economic Advisors, 1998; National Center for Health Statistics, 1998b). In addition, many communities contain potential sources of environmental health risks (e.g., industrial facilities, waste treatment sites, or waste disposal sites). These can affect all racial, ethnic, and socioeconomic groups, but there is substantial evidence that minorities and lower-income groups face higher levels of exposure to these hazards in terms of both frequency and magnitude. Although direct links between exposures and health are weak in many instances, the committee believes that allegations of environmental justice problems are frequently well founded and must be taken seriously enough to warrant careful assessment.

3

Research

As discussed in Chapter 2, there is substantial evidence that communities of concern bear disproportionate burdens of exposure to environmental hazards and the subsequent adverse health effects. Although environmental justice has many facets (e.g., legal, economic, and political), it may be approached appropriately in a variety of ways by the public and private sectors, and the health community should naturally focus on the health aspect of environmental justice. This aspect is most appropriately viewed as a public health issue—one for which public health perspectives and methodologies can contribute constructively to the clarification and resolution of the environmental health issues raised by the communities of concern about environmental justice. The committee recognizes that the diagnosis, treatment, and prevention of adverse health outcomes caused by environmental health hazards require a good understanding of the biological and physiologic mechanisms by which such hazards cause disease and that these mechanisms act separately as well as in combination. However, the larger community- and population-based issues of environmental justice require a public health perspective. A public health approach will contribute to environmental justice most effectively by examining issues on a broad, population basis, comprehensively identifying hazards to human health, carefully evaluating the adverse health effects of such hazards, developing alternative interventions to reduce or prevent risks, and evaluating such interventions rigorously to determine the most effective way to reduce risk and improve the health of the population (Institute of Medicine, 1988a).

A public health approach to environmental justice will also require a special relationship to the communities being studied, entailing unusual degrees of collaboration if research is to be responsive to the population's needs and if the findings are to be effectively implemented. As the existing literature in this area demonstrates, however, environmental justice research is an evolving and complex endeavor. This chapter describes the role of the public health sector in environmental health and the particular challenges facing environmental health sciences research in terms of both data and resources. The committee describes the current state of the research in environmental justice, discusses the research methodologies that best serve the communities of concern, and makes recommendations to improve current efforts to respond to environmental justice concerns.

RESEARCH METHODOLOGIES

Public health research on environmental justice issues incorporates two tasks: (1) assessment of the health status of the community and (2) determination of the contributions of specific environmental factors to that status. The public health community has developed a great deal of experience and competence in assessing the health status of the population. However, assessment of the health of racial or ethnic minorities, or low-income subpopulations in support of environmental justice poses difficult challenges because both the numbers of individuals and the incidence of disease may be quite small. Even greater challenges are posed by the second task—determination of the contributions of specific environmental factors. These challenges include documentation of excessive exposures, including their strengths and pathways; assessment of the susceptibilities of the communities of concern to environmental hazards; and measurement of the health effects of exposure, including the contribution of a specific hazard relative to the contributions of a variety of other potential factors. These analyses are also complicated by the problem of small numbers. The following section explores these environmental research challenges, with a strong emphasis throughout on the need for substantial involvement of the affected communities.

Documenting Excessive Exposures

A variety of sources of data might be useful for a public health assessment of a suspected environmental justice problem. The Environmental Protection Agency's (EPA's) *Inventory of Exposure-Related Data Systems Sponsored by Federal Agencies* (Environmental Protection Agency, 1992) lists 67 databases used by federal agencies to fulfill their responsibilities for research, regulation, and risk communication on environmental health issues (Environmental Protection Agency, 1992). The databases are managed by 17 federal agencies, the United Nations Environment Program, and the World Health Organization.

In addition, a number of private-sector databases focus on environmental health. Regulatory support is provided by 19 systems, and 29 systems focus on research. Twelve separate federal departments and agencies collect data relevant to the issue of environmental justice (see Table 3-1). Each has its own mandate and collects the data that meet its specific needs. Because the data are frequently received from local and state computer databases and are largely developed

TABLE 3-1 Selected Environmental Health Issues and Responsible Federal
Departments or Agencies

Environmental Health Issue	Responsible Federal Agency
Potential hazards, occupational or environmental, and accidental exposures	Department of Defense (DoD), Department of Energy (DoE), Department of Health and Human Services, Department of Labor, Department of Veterans Affairs, Environmental Protection Agency (EPA), Federal Emergency Management Agency
Manufacture, transportation, storage, and disposal of hazardous chemicals	Department of Commerce (DoC), DoD, DoE, Department of Transportation, Consumer Product Safety Commission
Exposure pathways (including air, water, and soil)	Department of Agriculture, Department of the Interior, DoC, and EPA

SOURCE: Institute of Medicine, 1997.

independently, few standardized methods for data collection, storage, analysis, retrieval, or reporting exist. The data gathered at the federal level are of various qualities and scopes (Council on Environmental Quality, 1993). The result is a patchwork of data that can be difficult to analyze comprehensively. Efforts are under way to coordinate federal environmental health data. These include the Department of Health and Human Services' Interagency Environmental Health Policy Committee and the National Environmental Data Index.

Examples of the various types of environmental health databases are discussed below. The Toxic Chemical Release Inventory (TRI) database is an example of a factual database based on geographic and industrial information. The databases that make up the Toxicology and Environmental Health Information Program (TEHIP) are either bibliographic (with citations and abstracts of the scientific literature) or factual (with data from scientific studies).

Toxic Chemical Release Inventory

Those concerned with environmental health often rely on EPA's TRI for information on the environmental releases of more than 300 toxic chemicals. Facilities are required to report emissions if, among other requirements, they process or manufacture more than 1,300 kilograms (25,000 pounds) of the chemical per year. Data in TRI include facility identification and the extent of environmental releases (including air emissions, water discharges, waste treatment, and releases to underground injection). The TRI database is particularly useful for community organizations in assessing local environmental hazards because it can be searched to identify facilities by zip code, city, county, or state. TRI is available through the National Library of Medicine.

Toxicology and Environmental Health Information Program

The mission of TEHIP, which is run by the National Library of Medicine, is to provide selected core toxicology and environmental health information resources and services, facilitate access to national and international toxicology and environmental health information resources, and strengthen the information infrastructure of toxicology and environmental health (National Library of Medicine, 1993). Each database has its beginning in several different federal agencies, which in and of itself leads to fragmentation of authority, responsibility, and accuracy. TEHIP is composed of 15 on-line environmental health databases that are developed or reviewed by the National Library of Medicine or another federal agency (including EPA, the National Cancer Institute, the National Institute of Environmental Health Sciences, and the National Institute for Occupational Safety and Health). Six of the databases are bibliographic; the other 10 databases provide factual data on chemical identification, carcinogenicity, mutagenicity, general toxicity and risk assessment, and environmental releases.

Implementation of a public health approach to environmental justice problems requires a solid research base and a reliable, comprehensive surveillance and reporting system. However, bibliographic and factual databases that can assist in research on and in understanding environmental health issues are available. Although substantial data are being collected, there are problems due to the lack of standardized definitions and methods and to the lack of standardized methods of data collection and retrieval, as well as significant gaps in the types of data that are needed to evaluate the effects that exposures to environmental hazards may have on the health of exposed populations. These shortcomings underscore the need, when undertaking an assessment of a potential environmental justice problem, to involve the affected community integrally in the process, particularly for the purpose of supplementing other sources of data with specific information pertinent to local conditions. Strategies and issues regarding such community involvement are discussed more fully later in this chapter.

Assessing Susceptibility to Environmental Hazards

When examining environmental justice from a public health perspective, it is important to recognize that communities of concern might be disproportionately affected not only because of their higher levels of exposure to environmental hazards but also because, for a variety of reasons, such exposures have a greater effect on them than on other communities. It is therefore important to examine potential differences in the susceptibilities of members of these communities to adverse health effects. Rios and colleagues (1993) have reviewed inborn and acquired variations among minority populations in their susceptibilities to the effects of environmental exposures. They reported that susceptibility can be affected by genetic factors (e.g., the sickle cell trait may increase one's susceptibility to the toxic effects of carbon monoxide), dietary factors (e.g., lower calcium intake among African American children may act to increase gastrointestinal absorption of ingested lead), other lifestyle factors (e.g., smoking increases lung cancer susceptibility in asbestos-exposed workers), or other environmental exposures (e.g., concurrent solvent exposure may increase the likelihood of hearing loss due to high levels of noise) that may be associated with variations among minority populations. Additional factors that Rios and colleagues concluded may differentially affect minority populations include compromised health status (e.g., people with diabetes may be less able to detoxify organic solvents), social inequality of access to health care (e.g., poor control of asthma by primary care providers may increase susceptibility to particulate air pollution), and inadequate education and communication skills (e.g., non-English-speaking workers may not be able to read health and safety warnings at work).

Frumkin and Walker (1997) also reviewed some of the mechanisms that act to increase the risk of environmental and occupational diseases among minority workers and communities. In addition to disparities in exposures and susceptibilities to environmental agents in the community and workplace, the investigators pointed out that the racial or ethnic and socioeconomic disparities that exist in access to health care in general may contribute to observed differences in occupational and environmental illnesses, although further research is needed to clarify this.

One tool that can be used to identify increased susceptibility is biomarkers. Biomarkers are measurements of the body's response to external events or substances such as environmental hazards. A biomarker of susceptibility would measure limitations, either inherited or acquired, in a person's ability to mount a protective response to a hazard. The development of biomarkers of susceptibility would allow further analysis of differentials in susceptibility among minority or, possibly, low-income populations. A more complete discussion of biomarkers is provided later in this chapter.

Measuring the Health Effects of Exposure to Environmental Health Hazards

Establishment of the causal relationship between exposure to environmental hazards and adverse health outcomes and measurement of the scope and severity of such outcomes are critical steps in the analysis of environmental justice issues. As noted at the beginning of this chapter, a better understanding of the disease mechanisms and the processes involved is needed. The committee's principal focus, however, is the use of epidemiologic studies in communities of concern. Such studies are designed to discern relationships between health effects and potential causes. In his recent review of environmental justice, Foreman asserts, "For environmental justice to contribute measurably to public health in low-income and minority communities, it would almost certainly have to stress an epidemiologic perspective . . . to a far greater extent than is currently the case" (Foreman, 1998, p. 70). Here, again, much of the research to date has been undertaken in occupational health. Two of the biggest challenges to an epidemiologic analysis of health effects are the existence of multiple exposures in the community of interest and the possibility that an adverse health outcome may have multiple determinants.

Multiplicity of Hazards

A community of concern may be exposed to multiple environmental hazards, which may act cumulatively or which may even interact in complex ways to magnify their risks to human health. Researchers have long recognized the need for knowledge about the separate and collective health effects of multiple chemicals (National Research Council, 1988). Recent research into the risks from mixtures has begun to provide some insight into the toxicological issues underlying the interactions of chemical and physical agents and the interactions of external exposures and chemotherapeutic drugs, as discussed in the recent report *Interactions of Drugs, Biologics, and Chemicals in U.S. Military Forces* (Institute of Medicine, 1996).

Research on multiple exposures has tended to focus on the occupational setting. It is important for new research, however, to take other settings into account, especially residential settings (see Box 3-1). Although it will never be feasible to study the health effects of every specific mixture that may occur because the possible number of combinations is very large, it should not be difficult to select those combinations that should be given priority because of their larger concentrations or greater likelihood of having toxic effects.

Multiplicity of Potential Determinants

A given health condition identified in a community of concern may have several possible etiologies. A major challenge for conducting research on a population with a high prevalence of diseases with multiple causes, such as asthma, is to identify the important environmental determinants from the multiple other factors that affect that disease's expression in populations or individuals. Current models of exposure assessment do not adequately capture this type of interaction. Existing health or environmental exposure databases do not include information on all the relevant factors that need to be investigated to answer many of the research questions posed. Allergens, for example, are notably absent from the majority of environmental exposure databases. Even if a

---

BOX 3-1

Altgeld Garden, Chicago

Altgeld Garden is a public housing community in southeast Chicago built in the mid-1940s. Since its construction, it has been surrounded by industrial facilities. Initially, these were heavy industry (steel, petrochemical, and manufacturing). Currently, they include manufacturing and water and waste treatment facilities. In the surrounding area are more than 100 industrial plants and 50 active or closed waste dumps. The area contains 90 percent of the city's landfills.

As a consequence, the 10,000 residents of Altgeld, who are predominantly African American, have been exposed to a broad variety of environmental stressors, raising concerns about the impacts of these exposures on their health. In addition to worrying about airborne exposures, many members of the community also believe that their houses are constructed on top of chemical and biological wastes. Well water has been found to contain cyanide, benzene, and toluene. According to EPA, this section of Chicago has the city's highest concentration of ambient lead and the second highest concentration of fine dust particles (Motavalli, 1998). Community health concerns have focused on several endpoints, with childhood cancer, prostate, bladder, and lung cancer, endocrine disease, hypertension, infant mortality, and asthma being the most prominent. In response to these concerns, several federal, state, and local agencies have evaluated or investigated the community.

---

single cytokine or other biomarker (see below) is eventually identified, for example, as a means of diagnosing asthma, it is unlikely that the environmental factors most responsible for the expression of asthma in communities of concern will be identifiable by a single biomarker of exposure.

Different exposures act by different biological mechanisms and will require different biomarkers. Even after a reliable assay for a biomarker of exposure, susceptibility, or biological effect is developed in the laboratory, its clinical and public health applications remain to be determined. For these reasons, among others, research into exposure assessment needs to be multidisciplinary.

Biomarkers

In the 1980s, the inability to link exposures to health outcomes by population studies and traditional methods for the classification of exposures led to the study of biological markers, or biomarkers, as possible tools for exploration of the effects of environmental exposures (Cullen and Redlich, 1995). As initially described by the National Research Council (NRC, 1989a,b), biomarkers, in the context of environmental health, are indicators of the effects of external exposure as manifest internally, in biological systems or samples. They reflect molecular and cellular alterations that occur as a disease begins and progresses (DeCaprio, 1997). According to the conceptual paradigm first proposed by NRC in 1987, these events (biomarkers) can be indicative of exposure, susceptibility, or effect (National Research Council, 1989b)(Table 3-2). (The designation of the status of the marker is sometimes subjective and may not be mutually exclusive [DeRosa et al., 1993].)

The value of biomarkers for epidemiologic studies of environmental justice lies in their potential to indicate that an exposure has occurred and to predict the likelihood of adverse health effects. To be able to relate events to a specific health effect, however, one needs to know which events are associated with which disease outcomes and the degree of that association. The application of biomarkers to environmental health research requires extensive research on disease mechanisms, which is the linking of exposure to hazards at various doses to the preclinical signs of disease (Henderson, 1995). The following sections describe the three types of biomarkers and the research needs for each type.

Biomarkers of Exposure

One of the first areas of focus in biomarker research was the determination and measurement of exposure. These efforts moved the environmental health field from estimates of external exposures to measurements of internal biological events (Cullen and Redlich, 1995). Much of this early work measured carcinogens at the molecular level, examining, for example, DNA adducts, sister chromatid exchanges, and micronuclei in epithelial tissue. Although these markers are routinely used as evidence of exposure, they might also be considered biomarkers of effect; that is, in some cases, they might also be predictive of potential adverse health effects (DeCaprio, 1997). Although these markers provide plausible dose-response models, they do not adequately identify exposures with unknown carcinogenicities.

TABLE 3-2 Types and Definitions of Biomarkers

Type	Definition
Exposure	An exogenous substance or its metabolite(s) or the product of an interaction between a xenobiotic agent and some target molecule or cell that is measured in a compartment within an organism
Susceptibility	An indicator of an inherent or acquired limitation of an organism's ability to respond to the challenge of exposure to a specific xenobiotic substance
Effect	A measurable biochemical, physiologic, or other alteration within an organism that, depending on the magnitude, can be recognized as an established or potential health impairment or disease

SOURCE: National Research Council, 1989b, p. 2.

Commonly measured pharmacokinetic values can also be used as markers of exposure, for example, the presence of parent compound or metabolites in exhaled breath, blood, or urine or the appearance of macromolecular adducts or their degradation products in urine. Some markers of chemical exposure, such as the hematological changes that accompany high levels of exposure to lead or benzene, have been measured for decades. As early as 1976, hemoglobin adducts were being used as internal dosimeters of exposure to ethylene oxide and were later used as internal exposure biomarkers for aromatic amines, nitrosamines, and polycyclic aromatic hydrocarbons (DeCaprio, 1997).

In essence, exposure biomarkers are useful for measurment of the actual absorbed dose and the extent of delivery of the exposure to the putative site. These measurements are superior to ambient monitoring and questionnaire data (DeCaprio, 1997). To understand the relationship of such markers to prior exposures, however, one must know the rates of formation and clearance of the marker and the factors that influence those rates. Because of safety concerns about determination of these rates in humans, historically these studies have been limited to those that use animal models (Henderson, 1995). Interspecies variations in absorption and uptake complicate extrapolation of results of studies with animal to human populations.

To address the issue of the nonspecificity of these biomarkers, more recent work has focused on early-effect markers, such as oncogenes and tumor suppressor genes. These markers not only serve as indicators of exposure, for example, in studies of aflatoxin and lung cancer in asbestos workers (Brandt-Raut et al., 1992; Hollstein et al., 1993), but they also provide insights into the mechanism of the disease process itself (Cullen and Redlich, 1995). However, the utility of these markers is limited as well because they cannot adequately account for variability in individual susceptibility factors; that is, the dose-response curve differs among individuals due to differences in metabolic pathways. Still, better biomarkers of exposure could be very useful in verifying claims of environmental exposure in communities of concern.

Biomarkers of Susceptibility

A biomarker of susceptibility is an indicator of an inherent or acquired limitation of an individual's ability to respond to the challenge of exposure to an environmental hazard. The variation in individual responses to environmental exposures is wide, even within racial or ethnic classifications (see discussion in Addressing Race below). Investigtors have studied an extensive range of enzymes that are known to be important toxicologically and that also demonstrate substantial variation in activity levels within the population (e.g., N-acetyltransferase or P-450 cytochromes). Such enzymes are likely to play an essential role in the activation or detoxification of potent carcinogens or other chemical exposures. Different susceptibilities are likely to account for at least some of the different responses to exposures such as metals, solvents, or pesticides (Bock,

1992). A deeper understanding of susceptibility and the biomarkers that indicate heightened susceptibility would be a valuable tool in preventing avoidable adverse health effects due to environmental exposures to health hazards.

Biomarkers of Effect

The 1989 NRC report defined a biomarker of effect as "any change that is qualitatively or quantitatively predictive of health impairment or potential impairment resulting from exposures" (National Research Council, 1989b). Although these markers are more predictive of ultimate toxicity, they are less clearly associated with exposure to specific chemical agents (DeCaprio, 1997). That is, the presence of such a marker can be indicative of more than one exposure.

Some mutational events can be considered biomarkers of effect, especially if they have already been demonstrated to be the immediate precursors of clinical disease, for example, oncogene activation and tumor formation. In some cases, however, the distinction between a biomarker of exposure and one of effect is not clear. Thus, DNA adduct formation (biomarker of exposure) might or might not lead to subsequent mutations that are precursors of disease.

Limitations and Potential of Biomarkers in Environmental Health Risk Assessment

Biomarkers measure events along the continuum from exposure to effect. They are signal events but are not necessarily an explanation for an underlying pathophysiology. Nevertheless, they can be tremendously useful in environmental epidemiology. More work is needed to develop biomarkers of exposure and effect suitable for improving the power of epidemiologic studies, including molecular epidemiology. For example, biomarker screening studies could be conducted with residents living in close proximity to a site to provide information on the actual levels of uptake of the contaminant(s) of concern.

Large, collaborative research efforts that use batteries of biomarkers are needed. Markers selected for use in the screening of populations must be sensitive, specific, predictive, and selective (DeRosa et al., 1993). Selectivity refers to the ability to unequivocally identify a specific substance to which an individual is exposed. For example, urine phenol levels can be influenced by the ingestion of vegetables, exposures to several aromatic compounds, ingestion of ethanol, and inhalation of cigarette smoke; thus, their value as a selective marker is low (DeRosa et al., 1993).

Markers must also be sensitive to short- versus long-term exposure. For example, the presence of trichloroethanol (a short-half-life metabolite of trichloroethylene [TCE]) in urine is a good biomarker for use in the monitoring of populations after short-term exposure to TCE, whereas the presence of trichloroacetic acid (a long half-life metabolite) in urine would be a more appropriate marker for use in the monitoring of populations after long-term exposure (DeRosa et al., 1993). Of particular value would be markers that improve the attribution of disease endpoints to causes (e.g., allowing determination of which specific mutations in lung cancer tissue are a fingerprint for which environmental cause or which asthma attack is due to an industrial pollutant rather than a natural pollen).

For biomarkers to be useful in characterizing human health effects associated with environmental hazards, they must be validated with human populations (National Research Council, 1989b, 1992a,b). This will generally require a clinical research setting and the use of epidemiologic and multivariable biostatistical methods that adjust for the important potential confounding and effect-modifying factors that could be masking a real environmental effect or causing spuriously positive results. Improved epidemiologic and clinical research methods that can better distinguish truly harmful from harmless environmental exposures in humans, that can detect lower doses, and that require smaller sample sizes need to be developed and validated.

A challenge that faces researchers is the need to link biomarkers to the disease with which they are associated and to determine at what levels disease is induced. Two strategies should be used to determine the link between biomarkers of exposure and an individual's prior exposures. The first strategy, according to Henderson (1998), is physiologically based toxicokinetic modeling. In order to develop such a model, data regarding the rate of formation of a biomarker following exposure to a toxic agent and its rate of removal or repair in the body (e.g., rate of excretion or degradation half-life) must be obtained. Physiologic parameters (e.g., cardiac output or breathing rate), as well as the physical-chemical characteristics of the chemical and its metabolites must be determined. Second, multiple biomarkers can be used to elucidate prior exposures more in depth than what may be obtained with a single biomarker. For example, if both the amount and the half-life of a biomarker vary, this knowledge can be used to give more perspective on a previous exposure. This knowledge will be able to determine whether an individual was recently exposed to a high level of a chemical or was continuously exposed to low levels in the past.

The committee supports the view that one of the most promising ways to accomplish this is by incorporating appropriate biomarkers to improve the accuracy of measurement of exposures, susceptibility factors, or disease outcomes in well-designed epidemiologic studies (Hulka et al., 1990; Schulte and Perera, 1993).

RESEARCH CHALLENGES

Current research in toxicology, epidemiology, molecular biology, clinical medicine, and social sciences can make important contributions to the study of environmental health but cannot adequately address the range of issues raised by environmental justice. Collaborative approaches to research that incorporate these and other relevant disciplines should be developed to address specific public health problems. It is important that research concerning environmental justice use both traditional and nontraditional methodologies to best serve the communities of concern. These include creating new epidemiologic tools to assess small populations better, addressing race and relevant socioeconomic considerations in the analysis, and involving the community in every stage of the research through participatory research.

Improving Epidemiologic Studies

Questions of environmental justice tend to be raised on behalf of relatively small populations. Many of the problems associated with the study of small populations, such as minority and economically disadvantaged individuals, have to do with the fact that study size requirements, (in addition to other requirements, such as isolation of exposures) for traditional epidemiologic studies can rarely be met. Populations that are (or that are believed to be) dealing with illness as a result of a high level of exposure to environmental health hazards are often isolated, either in urban or in rural areas, and are typically subject to other factors that affect health and well-being. Moreover, environmental and health data for populations whose health may be affected because of exposure to environmental health hazards are not routinely collected or analyzed by demographic categories (Environmental Protection Agency, 1992).

Various socioeconomic factors are associated with different rates of morbidity and mortality among different racial or ethnic groups (National Center for Health Statistics, 1998b; Warren, 1993). To the extent that they are correlated with environmental exposures, these socioeconomic factors could be considered confounders (in the epidemiologic sense) of the relationships between the environmental exposures and the disease outcomes. Typically, even in the event that they are measured in environmental epidemiologic studies, the studies' designers attempt to control for these potential "confounders" statistically to assess whether an association exists between the disease and the exposure variables of primary interest.

Several investigators have commented on reasons for the paucity of data on the roles of race, ethnicity, and other socioeconomic factors in studies of occupational diseases (Friedman-Jiménez, 1994; Kipen et al., 1991; Zahm et al., 1994). Some research deliberately excludes analysis of differential disease occurrences in minority workers because the small number of minority subjects would have provided an unacceptably low statistical power to test the primary hypotheses of the study. This is unfortunate because the small body of published occupational health literature that does explicitly include data from studies with these populations suggests that racial, ethnic, and economic disparities continue to influence the risk for adverse health effects due to environmental hazards in the workplace.

Research is needed to improve the capacity of epidemiologic studies to detect adverse health impacts in small populations and to evaluate clusters of effects observed in populations. (Appendix A discusses these and other related issues in greater depth.) Researchers need to design environmental health studies that will provide adequate measurement, classification, and reporting of data on race, ethnicity, and relevant socioeconomic variables and to develop improved methods of descriptive, analytic, clinical, and molecular epidemiology that are accurate and practical for investigating relationships between environmental exposures and disease in low-income and minority populations.

One tool that can help epidemiologic studies is geographic information systems (GISs). Geographic data can be used to relate the location of a known or a suspected environmental health hazard to public health trends and racial distributions, among other factors. Because GISs can provide powerful summaries of relationships that may be lost in numerical analyses, they have been found to provide clues to relationships that can then be investigated by quantitative techniques (Elliott et al., 1996). Such techniques can also merge environmental and public health data collected from many different sources.

Addressing Race

Because a central focus of environmental justice is on disparities among racial groups, it is important that studies and research take account of race and socioeconomic factors. However, the committee took cognizance of concerns being raised about conventional definitions and classifications.

The use of racial and ethnic categories for health surveillance is often confounded with other differences, such as geography, economic status, culture, lifestyle, or behavior. In his synopsis of a workshop, Health Surveillance and Communities of Color, sponsored by the Centers for Disease Control and Prevention and the Agency for Toxic Substances and Disease Registry, Rabin (1994) stated that surveillance needs to "pay more careful attention to differentiation within minority populations regarding year of migration, family status, income, age, daily work habits, religion, [and] media habits" (p. 45).

The categorization of individuals simply by race ignores other variables that can lead to valuable insights into predictors of risk. Many groups, including the American Anthropological Association and the Institute of Medicine (IOM) Committee on Cancer Research Among Minorities and the Medically Underserved, have been critical of the use of the term "race" in health research, primarily because the term implies the existence of distinct human subgroups that differ fundamentally in biological makeup and origin and ignores the tremendous heterogeneity within such groups (American Anthropological Association, 1997; Institute of Medicine, 1999). In reality, "genetic diversity appears to be on a continuum, with no clear breaks delineating racial groups" (Marshall, 1998, p. 654).

The vast majority of health research on human population groups in the United States, however, has categorized populations according to familiar terms such as "white," "African American" or "black," "Hispanic," "Asian American," and other terms. Such categorizations may be reinforced by federal research agencies, such as the National Institutes of Health, which are required to comply with the U.S. Office of Management and Budget's guidelines for data collection with regard to U.S. population groups. The U.S. Office of Management and Budget's Directive 15 requires federal agencies to report population data on the basis of five "racial and ethnic" groups (U.S. Office of Management and Budget, 1978). The U.S. Office of Management and Budget notes that such classifications do not carry scientific or anthropological validity; rather, they are based on social and historical considerations. The American Anthropological Association and the IOM Committee on Cancer Research Among Minorities and the Medically Underserved note that this reporting requirement may handicap health researchers, who are often unable to draw meaningful inferences regarding the source of group differences because "racial" groups do not vary systematically with regard to biological or genetic makeup, socioeconomic status, culture, or other relevant variables.

The use of "race" in health research will be further complicated in the future because the U.S. census will allow respondents to list more than one "racial" category to describe themselves. Although it is expected that only a small fraction of the U.S. public will seek to describe themselves as belonging to more than one "racial" group (U.S. Bureau of the Census, 1997), health researchers will have to develop means of accounting for such populations.

Notwithstanding these concerns, the committee believes that the collection of race- and class-specific data is crucial if environmental justice is to be achieved. Data that lack specificity or that gloss over demographic and material realities will not support adequate analysis, regulatory intervention, or remediation of environmental health risks. In many instances, race is a critical variable with respect to environmental justice because (1) the racial segregation of neighborhoods remains a common feature of the United States, and (2) lower-income, predominantly minority neighborhoods may be especially vulnerable to environmental degradation and abuse primarily because of relatively lower levels of political and economic clout of the populations in those neighborhoods to gain redress or restitution from environmental polluters. "Race" therefore remains an important term because of its social and political implications, but it should not be assumed to have scientific validity in the absence of evidence.

Participatory Research

To better understand the consequences of exposure to environmental hazards, public health officials and researchers should pay close attention to the experiences of individuals in local communities and should systematically collect and validate data on those experiences. Adverse health effects from environmental hazards are often suspected first by the people who experience them rather than by the health care or scientific community. In other cases, a toxicant's effects may be known to the health or science professionals but particular routes of exposure may have yet to be discovered for a particular community. In still other cases, affected individuals may be the initial source of knowledge about multiple exposures or confounding physical conditions (e.g., compromised health because of disease or nutritional problems, as is observed in iron deficiency and lead absorption). In all of these cases, these individuals are first-hand observers, with unique and essential knowledge about the activities or places that may lead to exposure.

An organized, methodical system of collecting these experiential data is an essential part of the scientific process (National Research Council, 1991b, 1996, 1997). Frequently, affected individuals bear the burden of proof for establishing the legitimacy of their problems (see for example, Box 3-2). Without the assistance of those with scientific training and financial resources, this can be an impossible task. If early data are ignored pending conclusive confirmation, however, there is a risk of presuming a hazard to be safe on the basis of inadequate data, thereby subjecting exposed people to unnecessary harm.

Public health officials and researchers need to develop ways to help community activists and local medical personnel document health outcomes and health status in a reliable and unbiased manner. Even if the methods are imperfect, they could produce evidence for the justification of more thorough medical surveillance and measurement. Epidemiologic data obtained by laypeople cannot supplant data obtained by epidemiology professionals; they can, however, help identify issues or supplement the data obtained by professionals (National Research Council, 1991b, 1997).

One of the best methods for capitalizing on local knowledge is participatory research. Participatory research has been defined as research that involves the affected community in the planning, implementation, evaluation, and dissemination of results (Banner et al., 1995; Drevdahl, 1995). In this regard, scientists serve as a resource to the community and work with the community in identifying and finding solutions to environmentally related health problems. In addition to allowing researchers to capitalize on local knowledge, the involvement of the affected community ensures that the research addresses the issues that are important to the community and reinforces the social validity of "the goals, procedures, and effects" of the research—that is, participatory research ensures that the community truly benefits from the research being done (Fawcett, 1991, p. 235). Although participatory research can greatly benefit and advance environmental health sciences, it still poses unique challenges.

Challenges of Participatory Research

Cultural differences between a minority community exposed to an environmental health hazard and the majority of Americans may be a barrier to communication and may affect the collection of data and the understanding of the relationship between exposure and disease. For example, the American Indian and Alaskan Native communities hold as sacred free-ranging animals, wild herbs, and other flora and fauna, values that have resulted in the discussion of endan-

BOX 3-2

Tucson, Arizona

In 1951 a group of companies began aircraft maintenance operations around the Tucson Airport in Arizona, using a variety of toxic solvents, including trichloroethylene (TCE). The occupational medicine literature of the 1970s and 1980s also documented acute episodes of neurotoxicities (e.g., dizziness, confusion, memory loss, blurred vision, elevated anxiety, headache, and reduced visuospatial relations and psychomotor speed) among workers who received short-term exposures to high concentrations of TCE vapors.

For approximately 30 years, the companies in Tucson stored the used solvents, including an estimated 7,570 liters (2,000 gallons) of TCE, in large evaporative pools, from which they gradually leaked into the aquifer of the Santa Cruz River, the principal source of water for the city. Over the same period, new industries and low-cost housing developments spread throughout southern Tucson, adjacent to the airport. An estimated 90 percent of the new residents of southern Tucson were low-income Mexican Americans. The drinking water for this high-growth area was drawn from wells that tapped directly into the contaminated plume spreading within the Santa Cruz River aquifer.

In 1981, the Arizona Department of Health Services tested the wells for contamination and discovered that several municipal wells contained levels of TCE above the level requiring state action (5.0 μg/liter), ranging from 1.1 to 239 μg/liter (Agency for Toxic Substances and Disease Registry, 1996). The affected wells were immediately shut down, as were others that were later discovered through continued monitoring to have become contaminated. Still, the residents had been exposed to indeterminable concentrations of TCE.

It was not until Jane Kaye described higher rates of disease in the *Arizona Daily Star* in May 1985 that there was a much heightened awareness of the contamination and speculation of a link to health effects. Similarly, Carol Roos, a school district social worker, investigated the area's seemingly high rates of disease by spending 4 months in house-to-house "shoe-leather epidemiology" (Brown, 1993; National Research Council, 1991a,b). These articles had the effect of creating intense community concerns about disease and a desire for information, health studies, and environmental cleanup. Despite their scientific shortcomings, both studies had the effect of raising the public's concern about a potential link of TCE to adverse health events. Although no consistent link between disease (i.e., mortality rates of all diseases and incidence rates of birth defects, childhood cancers, and testicular cancer) and the contaminated area was found in the studies subsequently conducted by government agencies, it was the community's efforts that raised the visibility of the problems in the area and created a climate that drove all parties to seek answers and solutions.

gered wild foods not only in terms of environmental health but also in terms of a violation of cultural mores. Unfortunately, the public health sector often does not include trained personnel from these communities of concern, nor is it rich in individuals from the larger community who are able to understand, empathize, and sympathetically work in the cultural context of small communities. The committee's site visit to the Hanford Nuclear Reservation served as an example of the need to consider cultural differences among populations that might not only affect the extent of environmental exposure to toxic compounds but also influence policies toward public health research and policy (see Box 3-3).

BOX 3-3

Hanford Nuclear Reservation

In 1943, the federal government acquired 1,450 square kilometers (560 square miles) on the Columbia River in south central Washington State to use as a site for the production of plutonium for use in nuclear weapons. After it was discovered that underground tanks containing toxic radioactive waste from the Hanford Nuclear Reservation had leaked, allowing the waste to enter the Columbia River and the local groundwater (Harden, 1996; Zorpette, 1996), the federal government attempted to determine the level of exposure of the local population to the toxic material. According to the Hanford Dose Reconstruction Project, Hanford's releases resulted in low whole-body doses. Those living near Hanford before 1960 may have received high doses of radiation. Unfortunately, the initial dose reconstructions did not consider the American Indian population separately.

American Indian people live on a number of reservations located in this region, as close as 80 kilometers (50 miles) to the Hanford site and at distances of up to several hundred miles. The Yakima Indian Nation is the largest tribe with an interest in Hanford. Tribal lands were directly ceded to the Hanford Reservation. The Confederated Tribes of the Umatilla Indian Reservation and the Nez Perce also directly ceded land to form the Hanford Reservation. Fishing from the Columbia River, hunting, and gathering remain a central part of these cultures and economies; thus, environmental contamination would have more than an adverse health impact. Historically, the Nez Perce lived outdoors, in camps, and moved around the land seasonally. They were not informed of the contamination at Hanford, potentially placing them at greater risk than the non-Indian population (Nez Perce Tribe, 1995).

Later dose reconstructions described to the committee have taken special account of the diet and societal tradition of these people and provide a template for consideration of individual subpopulations in the dose reconstruction and prediction of levels of exposure to radioactive or chemical pollutants. Nevertheless, the complexities in filing, storing, and retrieving the myriad classified and unclassified documents associated with site activities pose significant challenges to tracing the information needed to assess human exposure to radiation.

Often, differences between the scientific and lay communities may also obstruct interactions between them. Typically, the two groups speak different languages, live in different places, and have different stakes in the proceedings. Their communication difficulties can lead to mistrust as well as misunderstanding. If the residents of an affected community distrust the researchers, for example, they may choose to withhold information. That distrust can come from the perceptions that the researchers are working against their interests, are advancing their own careers at the community's expense, or are simply disrespectful. Solicitation of community input prior to the beginning of the research, active community participation in the research implementation, and communication of results to the community could prevent these misperceptions.

Participatory research has been used in a variety of health-related areas (Banner et al., 1995; Cornwall and Jewkes, 1995; Drevdahl, 1995; Lillie-Blanton and Hoffman, 1995). There are inherent difficulties, however, in incorporating community input into scientific research. These difficulties can include the time required to undertake community participation, the need for an established community structure with which to work, and the process by which new knowledge is validated (Drevdahl, 1995).

Lillie-Blanton and Hoffman (1995) discussed five major areas that are important to building a mutually beneficial relationship between the scientist and the community. These areas include (1) scientist knowledge of the community, (2) the development of an appreciation for the policy and programmatic issues underlying the research, (3) clarification of the decisionmaking process, (4) the development of trust and respect, and (5) the development of community expertise and capacity.

The committee believes that public health research on environmental justice issues would be substantially improved by the development of one or more standard models of how best to undertake participatory research. The committee is also mindful, however, of the suggestion by Cornwall and Jewkes (1995) that the most important facet of participatory research lies not so much in methods but in researchers' attitudes.

Enhancing Support for Research

The foregoing discussion demonstrates that greater attention and greater resources are needed to improve the science base and research methodologies required to address environmental justice issues effectively. Such commitments are difficult, however, given the current status of research in this field. Environmental health and environmental justice are relatively young, emerging fields of inquiry. Consequently, it is often hard to marshall the resources warranted to pursue promising research. Large numbers of research scientists have not yet developed research proposals in these areas. Nor are large numbers of scientists who are knowledgeable about these fields or who consider them to be of high priority participating on review panels for research awards. In the highly competitive world of biomedical and public health research, the result is that research projects on environmental health and environmental justice represent a small proportion of the approved and funded projects. It also means that talented young researchers will be hesitant to assume the risk of committing to work that is considered unconventional and for which future funding is highly uncertain.

CONCLUSIONS AND RECOMMENDATIONS

Public health research will be particularly important to improving environmental health and achieving environmental justice. The committee believes that an epidemiologic approach should be the central means of dealing with the environmental health problems in disadvantaged communities. This approach is hindered at the present time, however, by the shortcomings in current databases and data collection methodologies. New research models and techniques are needed. Communities of concern must participate in the identification of problems needing research and in the design and implementation of research.

Recommendation 1. **A coordinated effort among federal, state, and local public health agencies is needed to improve the collection and coordination of environmental health information and to better link it to specific populations and communities of concern.**

• *Strategy 1.1* Expand efforts and resources for in-depth evaluations of health status and risk monitoring in communities of concern. These efforts should involve the members of the affected population in discussing and making decisions related to issues that may have adverse environmental effects on communities and making decisions related to the remediation of existing environmental health concerns.
• *Strategy 1.2* Develop longitudinal, communitywide, baseline health assessments that provide both reason and context for studies specific to the impact of the environment.
• *Strategy 1.3* Construct a reliable surveillance system that not only tracks health status (e.g., through the use of biomarkers) but that also signals disproportionate exposure.
• *Strategy 1.4* Include members of minority groups in research to better describe specific susceptibilities and health effects.
• *Strategy 1.5* Connect environmental exposure databases and up-to-date demographic data, including data on age, gender, race, ethnic background, employment, housing, educational attainment, and income.
• *Strategy 1.6* Build strong links between public health practitioners and the community's broader array of medical, dental, and nursing professionals to stimulate greater sharing of data and experience.
• *Strategy 1.7* Promote the wider distribution and dissemination of public environmental health databases.

Recommendation 2. **Public health research related to environmental justice should engender three principles: improve the science base, involve the affected population, and communicate the findings to all stakeholders (see Box 3-4).**

The following strategies are recommended as a means of achieving Recommendation 2.

Strategies for Improving the Science Base

• *Strategy 2.1* Develop improved biomarkers of exposure, susceptibility, and biological effect as well as improved exposure assessment technologies.
• *Strategy 2.2* Focus additional research on human susceptibility, both genetic and nongenetic, to environmental causes of disease.
• *Strategy 2.3* Allocate a portion of all environmental health sciences research portfolios to environmental justice issues.

Strategies for Involving the Affected Populations

• *Strategy 2.4* Develop and use effective models of community participation in the design and implementation of research on environmental health and environmental justice.
• *Strategy 2.5* Give high priority to participatory research when addressing the research needs of communities with environmental justice concerns.
• *Strategy 2.6* Involve the affected community in designing the protocol, collecting data, and disseminating the results of research on environmental justice issues.

Strategy for Communicating with Stakeholders

• *Strategy 2.7* Ensure that communities of concern have a full understanding of the purposes, methods, and results of any research done in their communities.

BOX 3-4
Three Principles for Public Health Research to
Address Environmental Justice Issues

1. **Improve the science base.** More research is needed to identify and verify environmental etiologies of disease and to develop and validate improved research methods.

2. **Involve the affected populations.** Citizens from the affected populations in communities of concern should be actively recruited to participate in the design and execution of research.

3. **Communicate the findings to all stakeholders.** Researchers should have open, two-way communication with communities of concern regarding the conduct and results of their research activities.

4

Education

The explosion of new scientific knowledge in recent decades has had two extraordinary and contradictory effects that bear upon issues of environmental justice. First, the public has enormous faith that most questions can be readily answered once scientists take the time to focus on them; second, and conversely, scientists themselves are keenly aware of their inability to keep pace with every new development in their own and related fields. This is often true for health care providers as well, who struggle to understand and stay current on issues such as the effects of the environment on health. Moreover, as described by participants in the committee's site visits, current efforts at educating health professionals and the public about the importance of the influences of the environment on human health do not seem adequate.

During the site visits, the committee was struck by the extent to which the citizens of communities of concern felt virtually defenseless against what they believed to be unreasonable and unfair environmental impacts. They felt that they had been abandoned by both the government and the local industry. When attention was drawn to their issues, they felt exploited by lawyers and others, who were seen as taking advantage of their situation. The irony is that both government and industry are now trying to establish more cooperative joint efforts with community-based groups to control or even reverse environmental degradation. Neither government nor industry, however, fully appreciates the deficit of knowledge about the local hazards under which the residents of those communities must live and work. Such ignorance can lead to both neglected and exaggerated problems.

An important part of the solution to the needs and problems that have been described involves building a well-informed community response. The goal, therefore, is not only to augment the knowledge of each participant in the process of identifying, correcting, and preventing environmental risks but also to link each participant in a network that can help the community work as a whole.

45

Building this network requires improvements in professional and public education. This chapter examines the educational issues in communities of concern, including the education of both health professionals and the general public, and presents an overarching recommendation with related strategies for addressing these issues.

HEALTH PROFESSIONAL EDUCATION

During the committee's site visits, the symptoms and health conditions that people commonly tended to attribute to environmental causes included respiratory disorders, skin rashes, hair loss, gastrointestinal stress, blurred vision, fetal abnormalities, blood disorders, and a variety of cancers. It seems clear that physicians in all medical specialties need to be alerted to the possible relationship between these types of complaints and environmental phenomena and that the means of diagnosis of environmental causes of disease and illness need to receive more attention within the community of health care providers. Addressing the problem of diseases caused by environmental and occupational exposures is an issue shared among all health professionals, and it is the major focus of environmental and occupational medicine (EOM). EOM training focuses on diagnosing and caring for people exposed to chemical, biological, and physical hazards in the workplace, home, and community environments. Thus, EOM is the medical discipline most appropriate for addressing health issues related to environmental and occupational exposures. Unfortunately, but as expected, the site visits revealed that primary health care providers in the community lack specific training in EOM and that community residents lack access to clinical EOM services (Institute of Medicine, 1991, 1995a).

Enhancing Health Professional Education

Certain irreducible minimums of education and training need to be established, particularly with reference to minority and low-income communities, if the trends of environmental degradation that have thus far prevailed in many communities are to be corrected and reversed and if any more communities are to be prevented from experiencing the same fate. These minimums are needed in many areas, including medical education, public health professional education, and nursing and allied health sciences.

Medical Education

Most medical students and residents receive very little training in environmental health or occupational medicine. In most medical schools, the average classroom time for occupational and environmental medicine combined is an estimated 6 hours. Even though national policy and scholarship support are weighted heavily and properly toward the specialty of primary care, physicians in training are already in their residencies before they receive any effective instruction in ways to include occupational and environmental data in their history taking. Yet, for many patients, the etiologies of their illnesses are rooted in the environments where they live, work, and play. Thus, it is important to ensure that primary care physicians can do the basics of EOM, such as taking an adequate occupational and environmental history, and that they know the signs and symptoms of the most common occupationally and environmentally related diseases. In a prior report (Institute of Medicine, 1988b) the Institute of Medicine (IOM) recommended a minimum standard of competency for primary care physicians: "At a minimum, all primary care physicians should be able to identify possible occupationally or environmentally induced conditions and make the appropriate referrals for follow-up" (p. 63). However, the present committee also recognizes that health professionals practicing in these communities may have difficulty finding appropriately trained colleagues to whom they can refer their patients.

Compounding the lack of knowledge is a lack of involvement of health care providers in efforts aimed at educating members of the community (Greenlick, 1992; Ozonoff, 1995). Local health care providers are largely unaware of the scientific nuances involved in addressing the issues of the human health impacts resulting from environmental and occupational exposures (Institute of Medicine, 1988b, 1995a). There is a tremendous need to educate health care providers about the potential environmental impacts and concerns and, most importantly, to involve these providers in the dissemination of accurate information about local environmental health issues.

Enhanced education in EOM in the medical school curriculum has been recommended by IOM for many years (Institute of Medicine, 1988b, 1995a,b), and a variety of innovative approaches have been developed to operationalize these recommendations. The most recent report from IOM (1995a) on this subject developed and described competency-based learning objectives for graduating medical school students (see Box 4-1). General agreement exists that EOM should be integrated into the routine practice of medicine (Institute of Medicine, 1988b, 1995a; Rom, 1997; Rosenstock and Cullen, 1994). The EOM content in the medical school curriculum has not grown rapidly, however, and sustaining an EOM content in the curriculum often requires the ongoing efforts of an energetic and enthusiastic "champion" of EOM in each medical school.

One of the most valuable tools in teaching EOM to clinicians is the use of a list of "sentinel health conditions" that clinicians are expected to know. These are diseases or pathophysiologic conditions that are potentially occupational or environmental in etiology and for which the possibility of an environmental etiology needs to be considered in the differential diagnosis (Kipen and Craner, 1992). The sentinel health condition is the clinical application of the "sentinel health event (occupational)" originally proposed by Rutstein and colleagues (1983) for use in public health surveillance. Some examples of sentinel health conditions include pulmonary fibrosis, bronchitis, lung cancer, mesothelioma, carpal tunnel syndrome, contact dermatitis, and recent-onset asthma. The sentinel health condition is one tool that can be used as an aid in recognizing patients with an environmentally or occupationally related disease. Identification of a patient with a sentinel health condition is not diagnostic however, and must be followed by an appropriate clinical evaluation to confirm or rule out the environmental or occupational diagnosis.

In addition to the need for enhanced EOM training for primary care physicians, previous publications (Castorina and Rosenstock, 1990; Institute of Medicine, 1988b) have clearly documented the shortage of physicians with specialty training in EOM. They have called for large increases in the numbers of physicians being trained in EOM to meet current and future needs for these specialists. This need is particularly pressing in health care organizations that provide medical care to communities of concern. Physicians committed to working in these communities need to be trained in EOM both at the primary care level and at the EOM specialist level. Particularly needed are EOM specialists who can provide informational and clinical support to the primary care physicians who serve these communities.

BOX 4-1
Competency-Based Learning Objectives for Medical Students

1. Graduating medical students should understand the influence of the environment and environmental agents on human health on the basis of knowledge of relevant epidemiologic, toxicologic, and exposure factors.

2. Graduating medical students should be able to recognize the signs, symptoms, diseases, and sources of exposure relating to common environmental agents and conditions.

3. Graduating medical students should be able to elicit an appropriately detailed environmental exposure history, including a work history, from all patients.

4. Graduating medical students should be able to identify and access the informational, clinical, and other resources available to help address patient and community environmental health problems and concerns.

5. Graduating medical students should be able to discuss environmental risks with their patients and provide understandable information about risk reduction strategies in ways that exhibit sensitivity to patients' health beliefs and concerns.

6. Graduating medical students should be able to understand the ethical and legal responsibilities of seeing patients with environmental and occupational health problems or concerns.

SOURCE: Institute of Medicine, 1995a.

Attracting physicians to a specialty in EOM can be difficult, however, because of the labor-intensive and time-consuming nature of the work. Providing effective care for a single patient with an environmentally or occupationally related disease can require the investment of many hours of taking a detailed exposure and medical history, searching the literature and critically evaluating epidemiologic and toxicologic studies, making difficult judgments about causal relations, discussing risk, negotiating with patients and employers on the basis of the physician's knowledge and understanding of the available evidence (often in an adversarial setting), interacting with government agencies, and possibly being required to prepare and deliver testimony in court or at a workers' compensation hearing.

Moreover, mechanisms for reimbursement for these activities within the general framework of clinical medicine are not adequate, whether in fee-for-service, managed care, or Medicaid and Medicare environments. Special reimbursement mechanisms, such as workers' compensation, rarely provide timely payment at a rate commensurate with the physician's investment of time and effort. Although a detailed discussion of reimbursement issues is outside the scope of this report, this is an important issue and is probably even more important for physicians who choose to practice in minority or low-income communities.

Finally, training the currently practicing local physicians about what hazardous substances exist in their communities and what diseases these substances can cause is an important need. State and local departments of health and environment should play a role in this, as should industry. The availability of free continuing medical education for physicians and other health professionals in conjunction with the work of public health departments will greatly enhance the ability of caregivers to protect the public.

Public Health Education

The accredited schools of public health have been in the process of redefining their roles and missions. This process has been driven in part by public funding. Funding for public health practice, research, and education has been neglected, whereas funding for basic research has expanded. While the neglect of funding for public health practice research and education has continued, the public health workforce that was trained for such positions as sanitarians, air and water analysts, soil testers, food inspectors, epidemiologists, and public health educators has been graying and shrinking. In its place has come a new generation of public health school graduates who consequently focus their careers on basic research rather than the once traditional track of community-based public health assessment, surveillance, and service. This trend causes the committee to be concerned that the public health needs of communities are not being adequately addressed. In addition, resources are often inadequate to support basic public health services as well as the training of public health professionals. Moreover, when additional research is needed to inform decisionmaking regarding public health actions, resources may not be available to support the research or the research that is being done may not be designed or conducted in a way that will provide the needed information. Enhanced funding for participatory research may be one effective approach to ensuring that the research addresses the public health needs of the community.

It seems evident that the nation's schools of public health need additional resources to effectively continue their role in protecting and enhancing the health of the U.S. population in the next century (Institute of Medicine, 1988a). As stated in IOM's 1988 report *The Future of Public Health:*

> Schools of public health should establish firm practice links with state and/or local public health agencies so that significantly more faculty members may undertake professional responsibilities in these agencies, conduct research there, and train students in such practice situations. Recruitment of faculty and admission of students should give appropriate weight to prior public health experience as well as to academic qualifications.
>
> Schools of public health should provide students an opportunity to learn the entire scope of public health practice, including environmental, educational, and personal health approaches to the solution of public health problems; the basic epidemiological and biostatistical techniques for analysis of those problems; and the political and management skills needed for leadership in public health. (Institute of Medicine, 1988a, pp. 157–158)

Nursing and Allied Health Sciences

Through its site visits, the committee became aware that public health nurses, nurse practitioners, social workers, and others were often responsible for providing the link between the residents of communities of concern and health care providers. As inspiring as those stories are, they also indicated not only the haphazard nature of those connections but also how few connections exist relative to the communities' needs.

Similar to the role that nurses play in the health care system in general, nurses play a critical role in these communities, as do other professionals. Nurses are often the first point of contact for people seeking health care, especially those in rural or underserved populations. They are also the largest group of health care providers in the United States (an estimated 2.2 million), and in occupational health practice, nurses outnumber physicians by six to one (Institute of Medicine, 1995b). For these reasons, among others, nurses should be trained in and should be familiar with environmentally related health conditions.

The education of nurses and other health and social service caregivers needs to reflect environmental concerns as well. This matter has been addressed in prior reports of IOM and the National Academy of Sciences, most recently, *Nursing, Health, and the Environment: Strengthening the Relationship to Improve the Public's Health* (see Box 4-2), a report of the Committee on Enhancing Environmental Health Content in Nursing Practice (Institute of Medicine, 1995b). Like nursing, all areas of health science need to make environmental health competence a priority in education and training.

BOX 4-2

General Environmental Health Competences for Nurses

I. *Basic Knowledge and Concepts*

All nurses should understand the scientific principles and underpinnings of the relationship between individuals or populations and the environment (including the work environment). This understanding includes the basic mechanisms and pathways of exposure to environmental health hazards, basic prevention and control strategies, the interdisciplinary nature of effective interventions, and the role of research.

II. *Assessment and Referral*

All nurses should be able to successfully complete an environmental health history, recognize potential environmental hazards and sentinel illnesses, and make appropriate referrals for conditions with probable environmental etiologies. An essential component of this is the ability to access and provide information to patients and communities and to locate referral sources.

III. *Advocacy, Ethics, and Risk Communication*

All nurses should be able to demonstrate knowledge of the role of advocacy (case and class), ethics, and risk communication in patient care and community intervention with respect to the potential adverse effects of the environment on health.

IV. *Legislation and Regulation*

All nurses should understand the policy framework and major pieces of legislation and regulations related to environmental health.

SOURCE: Institute of Medicine, 1995b.

Increasing the Number of Minority Health Professionals

In addition to the problems identified above, the community of health professionals seems in danger of losing its earlier zeal for recruiting more new professionals from among the country's racial and ethnic minorities. In many regions of the country where environmental insults have occurred (e.g., Tucson, Arizona), the presence of large numbers of minority health professionals could well have sensitized the residents of those affected communities as well as their political and business leadership and could have mitigated, if not prevented, the environmental damage. On the basis of the experiences of a variety of professionals, notably in the fields of public education, medicine, and law, increased participation by racial and ethnic minorities facilitates the delivery of vital services to ethnic and racial minority communities in the United States.

Increasing the number of individuals from the affected populations in the ranks of health professionals strengthens the tie between the profession and the community. Access to care can also be influenced by this. Individuals in many at-risk communities have inadequate medical care. An important approach to improving health care in the communities of concern would be to increase in medical schools the numbers of students who have demonstrated a commitment to providing medical care in such communities. Commitment to communities and to public health is probably more important than a preexisting interest in environmental medicine. Otherwise, unselected students who come from minority communities may be more likely than students from other communities to return to those communities to practice medicine (Moy, 1995). These students could be targeted for training in EOM at the medical student, house staff, and attending levels and would be more likely than students who went on to practice in other communities to have a significant impact in the affected communities. A targeted scholarship program might facilitate the recruitment process.

EDUCATION OF THE PUBLIC

Making the community aware of the environmental risks to its health is a vital step in combating morbidity and mortality. According to the participants in the committee's site visits, however, not only does the public receive little information about environmental health issues but the information that is presented is often delivered in language that is too technical or full of jargon and is illustrated with examples that are obscure or culturally insensitive. Moreover, the communities most at risk of a lack of environmental justice are frequently the least likely to receive information, and the information that they do receive tends to stop short of providing any guidance as to what the community can do on the basis of that information for its own protection. The site visit participants also reported that the information that is provided is typically prepared by and for nonminority communities and is seldom evaluated for its comprehensibility by the intended audience.

Many of the people who spoke with the committee during the site visits were optimistic about raising the level of knowledge of members of the community to a level where they could be effective participants in the creation of solutions. For example, the belief in community potential is the heart of the Community Outreach and Education Program (COEP) in New York City's South Bronx. Using "bidirectional communication," COEP engages "neighborhood residents . . . as partners in a collaboration that seeks to address their needs and concerns while finding answers to fundamental scientific questions" (Claudio, 1996b). In this way, the residents "become actual experts in their own neighborhood." Belief in the potential of people in the community is also the basis of the Ohio Outreach Strategy for Equity in Environmental Issues program, which is an outreach strategy committed to helping "inform, educate, and empower people of color on issues of environmental hazards that may exist in their communities" (King, 1996). The committee was also heartened by the description of the program called Responsible Care: A Public Commitment begun by the Chemical Manufacturers Association, a program that gives laypeople access to environmental information to help them become effective members of community advisory panels that will be involved in an interactive dialogue with the industry. These programs are good examples of the types of activities that would be helpful in all affected communities.

The challenge of educating and informing members of an affected community is complex, but the informed participation of citizens is needed to correct past injustices and prevent future ones. Addressing environmental justice concerns in this way will also serve to increase the human capital of the affected communities. Individuals who are better able to understand and deal with environmental justice issues should also be better able to deal with other challenges in their lives.

Education of the community should meet the following objectives: (1) make the community more aware of basic environmental health concepts, issues, and resources; (2) increase the role of the community in identifying problems related to environmental exposures; (3) involve the community in shaping potential research approaches to the problem; and (4) link community members who are (or who may be) directly affected by adverse environmental conditions with researchers and health care providers in developing and prioritizing responses. Methods of achieving these objectives include educating children, enhancing community leadership, and involving the community in research, that is, conducting participatory research.

Educating Children

The committee believes that children should be helped to understand, as early and as clearly as possible, how their health responds to external stimuli and what specific environmental hazards exist in their own communities. That information should be theirs by right, and they should be encouraged to become familiar with all real and potential environmental insults at the level of their own school, neighborhood, play area, and home.

The committee heard of many modest but successful efforts to bring environmental health and environmental justice concerns to the attention of kindergarten to 12th-grade students (Box 4-3 provides an example of such an activity). For example, the committee was impressed with Barbara Sattler's achievements with young people in Baltimore, Maryland. In a low-cost, low-key program of her design, Sattler significantly raised the young people's level of understanding of ambient noise, vehicular exhaust, industrial effluence, neighborhood degradation, and other discernible and measurable contaminants of their home environments.

Luz Claudio also reported on schoolchildren of the South Bronx, New York City, who sampled the Hudson River for evidence of polychlorinated biphenyls (toxic heat-transfer agents used as insulators in electrical equipment) and reported their findings to their communities. This activity is part of an overall Community Outreach and Education Program supported by National Institute of

---

BOX 4-3

**Teaching Environmental Health to
Elementary School Students**

ToxRAP (toxicology, risk assessment, and air pollution), a kindergarten to 8th-grade curriculum developed by Audrey Gotsch and colleagues at the Environmental and Occupational Health Sciences Institute at the National Institute of Environmental Health Sciences Center in New Jersey, uses air pollution examples to teach toxicology and environmental health risk assessment. In the 3rd- to 5th-grade curriculum, students learn a basic hazard assessment framework to guide their investigation concerning carbon monoxide poisoning by using a case study that describes a family who has been experiencing mysterious health symptoms. Following the class lessons, a teacher reported that one of her students made the link to her grandmother's complaints of headaches, stomachaches, and feelings of tiredness. On the basis of the information received from the granddaughter, the grandmother had her home checked and found that she had an elevated carbon monoxide level due to a faulty furnace. Several other teachers reported that their students had been successful in getting their families to buy carbon monoxide detectors; some students even requested carbon monoxide detectors for their holiday presents.

Environmental Health Sciences (NIEHS) and based at the Department of Community Medicine, Division of Occupational and Environmental Medicine, Mount Sinai Medical Center.

The committee also visited the Deep South Center for Environmental Justice, which trains at-risk youth for environmental careers. Building on its experiences, it should be possible to replicate such programs in other locales. For example, many chemical firms get involved with the communities where their facilities are located, and many others have expressed a desire to do so. Providing jobs for young people and participating in their training is one way to establish better contacts that benefit both the company and the community.

These examples are not the only ways of involving individuals from at-risk communities in environmental health and safety. Topics that capture the interest of environmental activists might be attractive candidates for conventional educational programs. The American Chemical Society (ACS) has attempted to increase interest in the chemistry profession by sponsoring a problem-oriented year-long high school chemistry course for college-bound students, Chemistry in the Community (American Chemical Society, 1998). In keeping with ACS's objectives, there is little social content to these lessons (although there is some treatment of risk issues). The imagination of students in communities of concern might be captured by curricula showing the beneficial and detrimental roles that chemicals play in their lives.

The committee believes, on the basis of these and many other, similar reports, that creative, innovative teachers can engage their students in the neighborhood-relevant and life-pertinent issues of environmental health and safety. In so doing, they would help prepare a generation of individuals who will be less vulnerable than their parents to acts that result in a lack of environmental justice. This will require some additional curriculum development.

Community Leadership

How effectively individual communities address environmental justice issues often seems to depend on the presence or absence of local leadership. The committee's site visits gave it the opportunity to meet with several local leaders (see Box 4-4; the committee is also aware of communities that have gone unorganized [Sampson et al., 1997]). From a social science perspective, it would be helpful to know what conditions encourage (or discourage) the emergence of community leaders so that similar situations might be created in other communities. From a practical perspective, one would like to create the conditions in which leaders emerge and work most effectively, focusing the community on its most critical problems. To the extent that they express the concerns of previously voiceless populations, these leaders might also be candidates for other public roles, which they might be trained to fill.

With technical issues, like those associated with environmental justice, lay leaders face the daunting task of becoming educated about complex and unfamiliar topics. They might have to learn the concepts and jargon of toxicology, epidemiology, risk assessment, regulatory procedure, and cost-benefit analysis. How well they succeed depends in part on the accessibility and effectiveness of technical support, training, and education. Various government and private organizations have been created to help them, and informal questioning during the committee's site visits suggested that the usefulness of these organizations varies considerably. As a matter of scientific and practical interest and importance, identification of the causes of success in these situations provides a special research opportunity for collaboration among social, physical, biomedical, and political scientists.

Conducting Participatory Research

As discussed in Chapter 3, the involvment of the members of a community of concern when conducting environmental health sciences research pertinent to environmental justice is essential for many reasons. One important result of participatory research is the educational opportunity for those members of a community of concern who participate. By becoming involved in the development, execution, and analysis of research that addresses their health concerns, they will become more familiar with the issues, the terms, and the methods of investigation. In addition, they will better understand the current state of the knowledge and its limits. This can be a very important means of encouraging participation, disseminating the results of research, encouraging community leaders, and generating trust in the research.

BOX 4-4

Living Is for Everyone, Nogales, Mexico

Nogales, Arizona, a community located on the U.S.-Mexico border, has fewer than 30,000 inhabitants. Population statistics are uncertain for Nogales, Mexico, immediately across the border, with estimates ranging from 200,000 to 350,000. A significant portion of the city on the Mexican side lacks electricity or running water. On the Mexican side, there is considerable industry, known as the *maquiladoras*.

A curtain of smoke or haze surrounds the city. The committee learned that this stems from discharges from manufacturing plants and from large-scale burning at dumps. Aside from these airborne transmissions of pollution, the people of Nogales live near an *arroyo*.* Emissions from a specialty resin plant nearby the arroyo send fumes to the areas where some of the largest clusters of systemic lupus erythematosus have been found.

Living Is for Everyone (LIFE) is a community-based clinic that began in 1991 when natives of the area shared their suspicions that an unusual number of residents were being diagnosed with serious diseases: systemic lupus erythematosus, multiple myeloma, leukemia, and other types of cancer. This led to a grassroots or lay epidemiologic study conducted chiefly by reviewing the death records of inhabitants of the city, noting the cause of death, and mapping areas where the deceased had resided. Death certificates were obtained from local funeral homes, and a map delineating the locations of diseased people was constructed.

The information obtained by these residents gained national attention. Some hoped that it would result in better health care; others worried about the bad image that it conveyed and the potential for damage to tourism. The committee was told that Arizona officials initially denied that there were any cancer cases in the area during the period because none appeared in the state registry. It was soon discovered, however, that the state's reporting system did not include Nogales. A later study by the University of Arizona confirmed the basic findings of the LIFE study, finding that the incidence of multiple myeloma from 1989 to 1993 was 2.4 times the expected rate; the systemic lupus erythematosus rate was 94.5 per 100,000 residents, compared with the highest published rate of 50.8 per 100,000. As a result, state and local governments and the local universities have provided support to LIFE and have become involved in monitoring the population.

*An *arroyo* is a small steep-sided watercourse or gulch with a nearly flat floor that is usually dry except after heavy rains. The *arroyo* near Nogales is filled with solids and runoff water from nearby industries and settlements from the Mexican side of the border.

A good example of participatory research that is applied to environmental justice issues is COEP, which was discussed earlier in this chapter. COEP is based in the Mount Sinai Environmental Health Sciences Center in New York City. The aims of this program are geared toward education: (1) increasing the numbers of students from underrepresented minority groups at the high school and college levels, health professionals, and workers; (2) serving as an information resource on environmental and occupational health problems; and (3) assisting the community in identifying and controlling toxic environmental hazards. A highlight of community-researcher collaboration in the South Bronx community served by COEP resulted in a proposal to NIEHS for funding for a research and intervention program that would address childhood asthma. The researchers responded to the questions posed by the community to "find what is in the air that is causing us to suffer respiratory problems and a high asthma rate" (Claudio, 1996a). The residents of the community will serve as study subjects but will also be involved in study design and implementation.

CONCLUSIONS AND RECOMMENDATIONS

Health professionals, community residents, and basic environmental health sciences researchers understand too little about environmental justice and environmental health issues. Building an effective community network that can identify, correct, and prevent environmental health risks requires enhanced efforts in the training of health professionals and education of the public. A collaborative community response to environmental health risks will help limit and prevent environmental insults and their harmful health effects. Such a community response requires that health professionals be able to diagnose environmentally related diseases, that the public understand the risks to community health, and that governmental and industrial leaders be responsive to the needs of the community. To this end, educational programs that will more effectively link all parts of the community and that will build a coherent network to meet the public needs need to be created or enhanced.

Recommendation 3. **The committee recommends that environmental justice in general and specific environmental hazards in particular be the focus of educational efforts to improve the understanding of these issues among community residents and health professionals, including medical, nursing, and public health practitioners. This would include the following:**

   • **enhancing health professionals' knowledge of environmental health and justice issues,**
   • **increasing the number of health professionals specializing in environmental and occupational medicine, and**
   • **improving the awareness and understanding of these issues by the general public.**

The following strategies are recommended as a means of achieving Recommendation 3.

Strategies for Enhancing Health Professionals' Knowledge of Environmental Health and Justice Issues

• *Strategy 3.1* Give environmental and occupational medicine a prominent role in the education of clinicians so that they recognize environmentally related diseases and integrate environmental medicine into their routine practice of medicine and can take a careful environmental and occupational exposure history.

• *Strategy 3.2* Educate primary health care providers and other health professionals about the issues of environmental justice as part of an enhanced curriculum in occupational and environmental health.

• *Strategy 3.3* Provide access for health care providers in communities of concern to relevant information and expertise in environmental and occupational medicine.

• *Strategy 3.4* Make state-mandated, federally certified copies of provider education materials and learning programs available for local care providers in all jurisdictions near sites where environmental toxicants are produced so that community health care providers are properly trained to respond to environmental hazards and accidents.

• *Strategy 3.5* Establish in state health departments and managed care organizations training programs in environmental medicine for practitioners in communities at risk. Continuing medical education workshops, fellowships, and other opportunities for clinical training should be made available as part of this activity.

Strategies for Increasing the Number of Health Professionals Specializing in Environmental and Occupational Medicine

• *Strategy 3.6* Develop academic, economic, and other incentives to help attract physicians and other health professionals into the practice of environmental and occupational medicine and health.

• *Strategy 3.7* Target academic awards and similar support for medical school faculty at all stages of their careers to environmental and occupational medicine specialists who have demonstrated a commitment to research, medical education, and clinical practice devoted to addressing the environmental and occupational health issues of low-income and minority workers and communities. In particular, this support should be targeted to the faculty of medical schools that train the primary care providers who will serve these communities and populations.

• *Strategy 3.8* Actively recruit to the public health professions individuals who are representative of those communities most affected or most at risk of environmental health problems.

• *Strategy 3.9* Recruit health professionals who have a demonstrated commitment to serving low-income or minority communities for education and training in environmental and occupational health. This training should include instruction in approaches to service delivery that are culturally appropriate. Incentives should be provided to the programs both to make the residents' clinical services accessible to the communities of concern and to implement and demonstrate the success of programmatic initiatives to increase the likelihood that the environmental and occupational medicine specialists will remain accessible to these communities after completion of their training.

Strategies for Community and Public Education

• *Strategy 3.10* Develop and establish interactive community education programs (e.g., conferences, videos, and town meetings) on environmental health issues, including possible hazards from subsistence foods (i.e., from hunting, fishing, and private gardens) in communities of concern. Local industry should be encouraged to be involved in these educational outreach programs. In all cases, cultural sensitivity and culturally appropriate communication should improve the effectiveness of the programs as well as the relationship with the community.

• *Strategy 3.11* Educate the general public about environmental justice issues. Outreach strategies could include the use of public service announcements, printed materials, and other effective outreach programs.

• *Strategy 3.12* Adopt state educational programs that include environmental health in the primary education of children, especially in areas of heightened risk.

5

Health Policy

Science should play a critical role in the formulation of public health policy. For a complex and newly developing subject such as environmental justice, however, science often cannot provide policymakers with research data pointing conclusively to a particular solution. Policymakers must therefore try to understand both what the science base is able to offer and its limitations. This understanding will help them weigh the merits of delaying a decision, in the hope of obtaining better data and analysis, or proceeding with a decision on the basis of imperfect knowledge and acknowledging uncertainties about its consequences. Decisionmakers who find themselves in this situation should also be very attentive to the inferences and presumptions with which they are operating; to what level they have assigned the burden of persuasion for or against a given option, and how it was chosen; and to ways of expanding not only the sources of additional information but also the processes and participants involved in arriving at a decision.

CURRENT SCIENCE BASE FOR
ENVIRONMENTAL JUSTICE

The committee believes there is ample evidence that racial and ethnic minorities and residents of low-income communities have a higher incidence of disease and a lower general health status than do majority and more affluent populations (National Center for Health Statistics, 1998b). In addition, the committee also believes there is substantial evidence that these populations also experience higher levels of exposure to potentially harmful environmental stressors (defined in this report to include noise, odors, and particulate matter, as well as toxicants, chemicals, and other pollutants) than other populations, although the committee acknowledges that the evidence is not consistent in either its methodological rigor or its conclusions (Brown, 1995; Szasz and Meuser, 1997). Beyond these beliefs, however, the chain of causality scientifically demonstrating a lack of environmental justice grows weaker.

One limitation is that with a few exceptions, an adequate body of knowledge directly establishing the link between those things researchers believe to be environmental hazards and specific adverse health outcomes typically does not exist. A good start on this research has been made in the occupational health field (Frumkin and Walker, 1997), but much more research is needed to document the nature, scope, and severity of the toxic effects of environmental stressors, particularly when multiple sources or types of potentially harmful substances exist. When causal relationships between environmental hazards and adverse health outcomes are established, risk assessments are needed to evaluate the relative contributions of the various hazards, both in terms of the numbers of people affected and the severity of illness. Judgments can then be made about research and intervention priorities.

A second limitation is that policymakers need better evidence about the precise roles of environmental hazards in creating adverse health outcomes and the burden borne by racial or ethnic minorities or demographically defined populations (Brown, 1995). Although evidence of disproportionate exposure to potentially harmful substances strongly suggests a disappropriate health burden, a variety of other factors might be involved. Clarifying these relationships will require detailed epidemiological studies, often on small, targeted populations, examining the relationships among disease incidence, environmental hazards, and other behaviors or exposures that may cause or exacerbate health problems. Properly disseminated, the research arising from such a broad-based public health perspective will help increase awareness that a given health outcome can be caused by multiple environmental hazards (with cumulative or interactive effects) and that a particular hazard can contribute to multiple adverse outcomes. Therefore, efforts are needed to evaluate total exposures and then to identify the most severe and remediable hazards as a guide to the most effective courses of action.

As discussed in Chapter 3, uncertainties in the scientific analysis of environmental justice issues have several sources, some reflecting limitations of current methodologies, others reflecting the lack of funding for implementing available methodologies. One large gap in current research data is the lack of tools for measuring important elements of environmental health and environmental justice issues, including health status indicators, sociodemographic characteristics, and other lifestyle factors that influence a population's health. Even where demographic and environmental databases do exist, it may be hard to link them. Small sample sizes and the inability to disaggregate data by such factors as race, ethnicity, income, and education further impede rigorous analysis and robust conclusions. In addition, it is very difficult to identify specific environmental hazards and to document their role in specific adverse health outcomes. These problems are even more difficult when there are complex interactions among multiple environmental hazards and long latency periods. Finally, little is known about the differential susceptibilities of various populations to environmental hazards. This suggests that greater attention, resources, and innovation are needed to develop the science base required to inform policymaking with respect to environmental health and justice.

Increased support for environmental health research may encounter political opposition from those who are concerned that the results of such research could increase their legal liability and the associated level of litigation they would face. Opposition to research and extreme skepticism about its results are rewarded if uncertainty leads to paralysis. It is the committee's view that policymakers need a philosophy that is consonant with the inherent difficulty of linking environmental hazards to adverse health outcomes. The populations in question generally have a lower health status than other groups in society and, therefore, are likely to be more susceptible than others to the adverse impacts of environmental hazards. Moreover, they are also likely to be less able to deal effectively with these hazards and their impacts, either in the political arena or by obtaining adequate access to high-quality health care.

It is a fundamental principle of statistical decision theory that the interpretation of scientific data requires making a value judgment about the relative importance of avoiding different kinds of errors. In the context of environmental decision-making, the possibility of error arises when human activity is suspected of causing ill-effects to people and their surroundings, but the linkage has not or cannot been proven. Society's response to such uncertainty reflects its willingness to tolerate one of two sorts of errors: (1) either failing to protect the public when there actually is a connection between an activity and a harmful result or (2) failing to protect the activity itself when it is not the source of the underlying problem. Which of these two types of errors is preferred is reflected in the manner in which the burden of proof is assigned. Are the activities presumed innocent until proven guilty (to some specified degree of confidence)? Or, rather, is the public presumed worthy of protection until the activities have established their innocence (with some specified degree of confidence)? There is no way of avoiding these choices.

This committee has concluded (as have others studying environmental justice) that social-political institutions tend to place the burden of proof on poor communities. This may not be done maliciously or even overtly. Rather, there are general procedures in place that foster inaction until scientific evidence of a firmer nature than most poor communities can muster has been collected. The procedures reflect, in part, the norms of basic scientific research, which place a premium on avoiding "false positives"—that is, accepting the existence of phenomena that do not actually exist. Although these norms have served the scientific community well, they place poor communities at a disadvantage. Such communities generally lack the ability to generate the scientific studies needed to confirm or disprove their suspicions, to examine the studies conducted by others, and to translate their concerns into terms that scientists employ. The committee urges the medical and scientific communities to make better use of lay observations, to make local communities partners in the design and conduct of scientific studies, to help citizens understand scientific data, and to help local public health officials and health practitioners deal with environmental health problems. If followed, the committee's recommendations should help establish a more equitable sharing of the burden of proof among the public, government agencies, and those engaged in potentially hazardous activities.

The committee also urges the scientific and political communities to recognize that the environmental problems faced by communities of concern are but one among many different problems they confront, including poor nutrition, limited education, minimal political representation, high unemployment, inadequate transportation, poor sanitation, hazardous workplaces, and insufficient vector control. When a community is under stress, primary attention should be given to attempting to address its health concerns in prudent ways. Secondarily, attention should be directed to identifying the accountable parties because holding people accountable for past misdeeds can be a valuable social function and provide incentives for appropriate behavior in the future.

EXPANDING THE POLICY PROCESS

As noted in prior chapters of this report, action, supported by adequate resources, is needed to develop and implement a public health strategy, to improve the science base, and to enhance awareness and understanding of these issues on the part of health professionals, educators, the business community, public officials, and the general public. In addition, policymakers need a more expansive perspective and approach to the development of public policy—and public-sector decisions—that have a bearing on environmental health and environmental justice.

Inevitably, in some situations delaying the decision-making process to wait for more data or better research is not a desirable or acceptable course. Delay may mean that a community will forego a significant economic benefit or may suffer further exposure to potentially dangerous environmental hazards. Foreman notes two other reservations about seeking solutions for environmental justice problems predominantly through scientific inquiry. He notes that "science cannot resolve what are ultimately value questions." In addition, he suggests that "calls for more and better scientific studies may, if successful, simply generate more information than policymaking institutions can reasonably digest given their available resources" (1998, p. 112).

These constraints on public policy-making do not mean that decisionmakers should forego evidence-based problem solving. To the contrary, it means that great attention must be paid to making sure that the decisionmaking process is as open as it reasonably can be to ideas, information, and participants derived from affected or potentially affected communities. In the case of existing environmental hazards, members of the local community may be the best available source of information regarding exposures—including multiple exposures or routes of exposure—or about the interplay of exposures and other potentially influential factors or susceptibilities that exist in the community. They may also be able to identify emerging patterns of illness before the health care delivery system or other possible sources can. Engaging the community not only has the advantage of increasing policymakers' knowledge and understanding of local problems, but can also give voice to community values. Hearing citizens' beliefs and concerns is essential to securing general acceptance of policies and actions. In this context, the community of concern needs to encompass all those likely to be significantly affected by a policy decision—including residents and community groups, business and labor, trade associations, and government and regulatory agencies.

Two case studies illustrate how effective stakeholder collaboration can lead to a broadly satisfactory resolution of a major environmental problem, and how its absence can result in stalemate and widespread dissatisfaction. In California, diverse interests were able to agree on long-range goals and strategies for preserving a valuable estuary (see Box 5-1). In Illinois, the attempt to impose a solution on a community that had been excluded from the decisionmaking process left the parties entangled in litigation that served environment and health policy needs poorly (see Box 5-2).

The Environmental Protection Agency's (EPA's) National Environmental Justice Advisory Council (1996; see Box 5-3) has developed a thoughtful and comprehensive plan, the Model Plan for Public Participation, which emphasizes planning and inclusiveness. Models and experiences from other areas of public policy are also available (National Research Council, 1996), but any model must be applied flexibly, must be adapted to specific local interests and concerns, and must be evaluated to ensure its effectiveness.

Policymakers should not shy away from the inclusion of community participants on the grounds that the issues are too technical or difficult for the general public to understand or respond to. The research literature generally finds that motivated lay audiences can understand many environmental issues, especially if a conscientious effort is made to determine their beliefs and concerns, and to develop appropriate communications (Fischhoff et al., 1997; National Research Council, 1989c, 1996). Even potentially esoteric material such as risk assessments or cost-benefit analyses can and should be understandable if presented in policy-relevant terms. When citizens object to the conclusions of an analysis, a careful examination is needed to see whether they fail to understand its premises or, instead, understand it all too well and object to how it summarizes existing studies, treats uncertainties, or defines risk (Fischhoff et al., 1981; Glickman and Gough, 1990).

CONCLUSIONS AND RECOMMENDATIONS

The committee concludes that concerns about environmental health and environmental justice are legitimate and should be taken seriously, even if the information and data related to these concerns still lack some of the rigorous scientific attributes that policymakers desire. Policymakers cannot assume that these concerns are without merit. However, policymakers should also recognize that many

---

BOX 5-1

**Successfully Engaging Stakeholders:
San Francisco Bay/Delta Accord**

Declaring "a major victory of consensus over confrontation" on December 14, 1994, California Governor Pete Wilson and Cabinet-level federal officials announced the signing of an historic agreement to protect the San Francisco Bay/Delta estuary—the largest and most productive estuary on the West Coast. Known as the Bay/Delta Accord, the agreement was negotiated by the leadership of the state's environmental, urban, and agricultural interests. The accord broke decades of gridlock on California water policy issues by establishing an integrated, ecosystem–based approach to protecting the estuary while providing more reliable supplies to the state's urban and agricultural water users.

The collaborative process that led to the accord marked a sharp departure from the decisionmaking approach traditionally used under the Clean Water Act and the Endangered Species Act. Rather than issuing proposals developed by individual agency experts for formal public comment and review, the agencies worked together with environmental, urban, and agricultural interests over 2 years to identify common goals and mutually acceptable solutions. The final standards were developed through an extensive peer-review process that involved both local and national experts in estuarine systems. This approach sharply reduced the number of legal and scientific challenges that accompany most major agency decisions and has been hailed as a national model for solving environmental problems.

Building on the success of this collaborative process, the state and federal agencies and interest groups have continued to work together as part of the new CALFED Bay/Delta Program to develop long-term ecosystem restoration goals. In 1996, the agencies and interest groups reached consensus on a $995 million bond measure that will help finance the ecosystem restoration process and other projects vital to the program's success. The bond was approved by voters in November 1996.

SOURCE: Presidential/Congressional Commission on Risk Assessment and Risk Management, 1997.

---

other considerations go into decisions such as choosing a site for a new manufacturing plant or solid waste facility, removing an alleged hazard, or imposing expensive environmental controls. Decisions of any significant consequence will almost always involve choices and trade-offs. Given the current state of knowledge, the committee believes that policymakers should be attentive to potential environmental hazards and adverse health outcomes and should be meticulous about including affected communities in the decisionmaking process.

Requiring residents of a potentially affected community to prove definitively that alleged adverse health outcomes are linked to environmental hazards may put effective participation beyond their means. Conversely, accepting assertions of environmental injustice without reservation could lead to actions that

---

BOX 5-2

**Insufficient Stakeholder Collaboration:
Granite City, Illinois**

When stakeholders are not included early in the decisionmaking process, they are more likely to oppose the risk management decision and block its implementation. This has been happening in Granite City, Illinois, since 1993, according to testimony from Mayor Ronald Selph and Alderman Craig Tarpoff. Heavily contaminated with lead by a former smelter, much of the city was designated by the Environmental Protection Agency (EPA) as a Superfund site. Based on soil sample analyses and a screening risk assessment model, EPA decided to remove the contaminated soil around 1,200 homes and businesses and haul it away.

Some believe that EPA made this decision without adequately consulting the community. City officials believe that this remedy ignored a number of problems, including health risks from dust and lead-based paint.

The industrial facility held responsible for the contamination did not respond to EPA's decision, so the agency sued the facility. The city then filed a petition in the suit because officials felt that neither EPA nor the responsible party represented the best interests of the community. EPA began the cleanup anyway but was restrained by court order. EPA retained an expert whose analysis supported the agency's choice of remedy, and the city retained an expert whose analysis concluded that the removal of contaminated soil would be fruitless unless the remaining sources of contamination—house paint, the smelter waste pile, and the trucking lot soil—were removed as well. Granite City residents are left confused and caught in the middle. Some support the city, and some support EPA. Property values have fallen. As of late 1996, the case remains unresolved and is back in federal courts.

SOURCE: Presidential/Congressional Commission on Risk Assessment and Risk Management, 1997.

---

adversely affect other interests without significantly improving a community's environment. Knowing the content and limits of the best available science will encourage the framing of reasonable inferences and help in determining the confidence to be placed in such inferences and the appropriate burden of rebuttal.

The committee believes that its recommendations are consistent with the collaborative view of risk management that has emerged from the series of National Research Council reports on this topic (National Research Council, 1983, 1989c, 1996b). The committee saw much goodwill in the communities it visited, which should allow for humane solutions to otherwise seemingly intractable environmental problems. The committee believes that lasting, workable arrangements can be created between communities of concern and others who have the capacity to use science and the law to help minority or disadvantaged populations protect themselves and their communities from harmful environmental stressors. Among those who will need to be involved in such arrangements are legislatures, regulatory agencies, business and industry, institutions of higher education, and the medical establishment.

Recommendation 4. **In instances in which the science base is incomplete with respect to environmental health and justice issues, the committee urges policymakers to exercise caution on behalf of the affected communities, particularly those that have the least access to medical, political, and economic resources, by taking reasonable precautions to safeguard against or minimize adverse health outcomes.**

BOX 5-3

Highlights from the National Environmental Justice Advisory Council's Public Participation Checklist

Use the following guiding principles in setting up all public meetings:

- Maintain honesty and integrity throughout the process;
- Recognize community and indigenous knowledge;
- Encourage active community participation; and
- Utilize cross-cultural formats and exchanges.

Identify external environmental justice stakeholders and provide opportunities to offer input into decisions that may affect their health, property values, and lifestyles.

Identify key individuals who can represent various stakeholder interests. Learn as much as possible about stakeholders and their concerns through personal consultation or phone or written contacts. Ensure that information-gathering techniques include modifications for minority and low-income communities (for example, consider language and cultural barriers, technical background, literacy, access to respondents, privacy issues, and preferred types of communications).

Solicit stakeholder involvement early in the policy-making process, beginning in the planning and development stages and continuing through implementation and oversight.

SOURCE: National Environmental Justice Advisory Council, 1996.

References

Advisory Board to the President's Initiative on Race. 1998. One America in the 21st Century: Forging a New Future. Washington, DC: U.S. Government Printing Office.

Agency for Toxic Substances and Disease Registry. 1996. Final Report: Disease and Symptom Prevalence Survey, Tucson International Airport Site, Tucson, Arizona. Atlanta: U.S. Department of Health and Human Services.

American Anthropological Association. 1997. Response to OMB Directive 15: Race and Ethnic Standards for Federal Statistics and Administrative Reporting. Arlington, Virginia: American Anthropological Association.

American Chemical Society. 1998. ACS Education: ChemCom [WWW document]. URL http://www.acs.org/education/currmats/chemcom.html (accessed October 23, 1998).

Anderton, D. L., A. B. Anderson, J. M. Oakes, and M. R. Fraser. 1994. Environmental equity: The demographics of dumping. Demography 31:229–248.

Baden, B., D. Grabowski, C. Vidos, and R. Yafchak. 1996. Infant health and hazardous waste in Chicago, 1990. Paper presented at the Workshop on Economic Policy and Public Finance, May 10, Chicago.

Banner, R. O., H. DeCambra, R. Enos, C. Gotay, O. W. Hammond, N. Hedlund, B. F. Issell, D. S. Matsunaga, and J. A. Tsark, 1995. A breast and cervical cancer project in a Native-Hawaiian community: Wai'anae cancer research project. Preventive Medicine 24:447–453.

Bock, K. W. 1992. Metabolic polymorphisms affecting activation of toxic and mutagenic arylamines. Trends in Pharmacological Science 13:223–226.

Brandt-Raut, P. W., S. Smith, K. Hemminki, H. Koshinen, H. Vainio, N. Niman, and J. Ford. 1992. Serum oncoproteins and growth factors in asbestosis and silicosis patients. International Journal of Cancer 50: 881–885.

Brody, D. J., J. L. Pirkle, R. A. Kramer, K. M. Flegal, T. D. Matte, E. W. Gunter, and D. C. Paschal. 1994. Blood lead levels in the U.S. population: Phase 1 of the Third National Health and Nutrition Examination Survey. Journal of the American Medical Association 272:277–283.

Brooks, S., M. Gochfeld, J. Herzstein, M. Schenker, and R. Jackson. 1995. Environmental Medicine. St. Louis: Mosby.

Brown, P. 1993. Popular epidemiology challenges the system. Environment 35(8):16–41.

69

Brown, P. 1995. Race, class, and environmental health: A review and systemization of the literature. Environmental Research 69:15–30.

Bullard, R. D. 1990. Dumping in Dixie: Race, Class and Environmental Quality. Boulder, CO.: Westview Press.

Carr, W., T. Zeitel, and K. Weiss. 1992. Variations in asthma hospitalizations and deaths in New York City. American Journal of Public Health 82:54–65.

Castorina, J. S., and L. Rosenstock. 1990. Physician shortage in occupational and environmental medicine. Annals of Internal Medicine 113:983–986.

Centers for Disease Control. 1991. Preventing Lead Poisoning in Young Children. Atlanta: U.S. Department of Health and Human Services.

Centers for Disease Control and Prevention. 1997. Update: Blood-lead levels—United States, 1991–1994. Morbidity and Mortality Weekly Report 46:141–146.

Claudio, L. 1996a. New asthma efforts in the Bronx. Environmental Health Perspectives 104:1028–1029.

Claudio, L. 1996b. Presentation to Institute of Medicine Committee on Environmental Justice, October 25, Chicago.

Cookson, W. O. C. M., and M. F. Moffatt. 1997. Asthma—An epidemic in the absence of infection? Science 275:41–42.

Council of Economic Advisors. 1998. Changing America: Indicators of Social and Economic Well-Being by Race and Hispanic Origin. Washington, DC: U.S. Government Printing Office.

Council on Environmental Quality. 1993. Environmental Quality: The Twenty-Fourth Annual Report of the Council on Environmental Quality. Washington, DC: Council on Environmental Quality.

Cullen, M. R., and C. A. Redlich. 1995. Significance of individual sensitivity to chemicals: Elucidation of host susceptibility by use of biomarkers in environmental health research. Clinical Chemistry 41(12 Pt. 2):1809–1813.

DeCaprio, A. P. 1997. Biomarkers: Coming of age for environmental health and risk assessment. Environmental Science and Technology 31:1837–1848.

DeRosa, C. T., Y. W. Stevens, J. D. Wilson, A. A. Ademoyero, S. D. Buchanan, W. Cibulas Jr., P. J. Duerksen-Hughes, M. M. Mumtaz, R. E. Neft, et al. 1993. The Agency for Toxic Substances and Disease Registry's role in development and application of biomarkers in public health practice. Toxicology and Industrial Health 9:979–994.

Drevdahl, D. 1995. Coming to voice: The power of emancipatory community interventions. Advances in Nursing Science 18:13–24.

Elliott, P., J. Cuzick, D. English, and R. Stern. 1996. Geographical and Environmental Epidemiology. Methods for Small Area Studies. New York: Oxford University Press.

Environmental Protection Agency. 1992. Inventory of Exposure-Related Data Systems Sponsored by Federal Agencies. Lexington, MA.: Eastern Research Group, Inc.

Environmental Protection Agency. 1993. Targeting Indoor Air Pollution: EPA's Approach and Progress. Washington, DC: U.S. Government Printing Office.

Environmental Protection Agency, Office of Federal Activities. 1998. Final Guidance for Incorporating Environmental Justice Concerns in EPA's NEPA Compliance Analyses. Washington, DC: U.S. Government Printing Office.

Fawcett, S. B. 1991. Social validity: A note on methodology. Journal of Applied Behavior Analysis 24:235–239.

Feinleib, M. 1993. Data needed for improving the health of minorities. Annals of Epidemiology 3:199–202.

Fischhoff, B., A. Bostrom, and M. J. Quadrel. 1997. Risk Perception and Communication, pp. 987–1002. *In:* Oxford Textbook of Public Health. R. Detels, J. McEwen, and G. Omenn, eds. London: Oxford University Press.

Fischhoff, B., S. Lichtenstein, P. Slovic, S. L. Derby, and R. L. Keeney. 1981. Acceptable Risk. New York: Cambridge University Press.

Foreman, C. H., Jr. 1998. The Promise and Peril of Environmental Justice. Washington, DC: Brookings Institution Press.

Friedman-Jiménez, G. 1994. Achieving environmental justice: The role of occupational health. Fordham Urban Law Journal 21:605–631.

Friedman-Jiménez, G., and L. Claudio. 1998. Environmental Justice. *In:* Textbook of Environmental and Occupational Medicine, 3rd ed. W. N. Rom, ed. Philadelphia: Lippincott-Raven.

Frumkin, H., and D. Walker. 1997. Minority workers and communities. *In:* Maxcy-Rosenau Public Health and Preventive Medicine, 14th ed. J. M. Last, ed. Norwalk, CT: Appleton-Century-Crofts.

Gergen, P. 1996. Editorial: Social class and asthma: Distinguishing between the disease and the diagnosis. American Journal of Public Health 86:1361–1362.

Glickman, T. S., and M. Gough, eds. 1990. Readings in Risk. Washington, DC: Resources for the Future.

Glickman, T. S., D. Golding, and R. Hersh. 1995. GIS-based environmental equity analysis: A case study of TRI facilities in the Pittsburgh area, pp. 95–114. *In:* Computer Supported Risk Management. G. E. G. Beroggi, and W. A. Wallace, eds. Dordrecht, The Netherlands: Kluwer Academic.

Goldman, B. A., and L. J. Fitton. 1994. Toxic Wastes and Race Revisited. Washington, DC: Center for Policy Alternatives, National Association for the Advancement of Colored People, and United Church of Christ Commission for Racial Justice.

Greenberg, M. 1993. Proving environmental inequity in siting locally unwanted land uses. Risk: Issues in Health and Safety 4:235–252.

Greenland, S., and J. Robins. 1994. Invited commentary: Ecologic studies—biases, misconceptions, and counterexamples. American Journal of Epidemiology 139:747–760.

Greenlick, M. R. 1992. Educating physicians for population-based clinical practice. Journal of the American Medical Association 267:1645–1648.

Harden, B. 1996. The dark side of paradise. Washington Post National Weekly Edition 13(37):6–10.

Henderson, R. F. 1995. Strategies for use of biological markers of exposure. Toxicology Letters 82–83:379–383.

Henderson, R. F. 1998. Finding a biomarker is a first step, pp. 33–39. *In:* Biomarkers: Medical and Workplace Applications. M. L. Mendelsohn, L. C. Mohr, and J. P. Peeters, eds. Washington, DC: John Henry Press.

Hollstein, M., C. P. Wild, F. Bleicher, S. Chutimataewin, C. C. Harris, P. Srivatanaki, and R. Montesano. 1993. P53 mutation and aflatoxin B1 exposure in hepatocellular carcinoma in patients from Thailand. International Journal of Cancer 53:51–55.

Hulka, B. S., T. C. Wilcosky, and J. D. Griffith. 1990. Biological Markers in Epidemiology. London: Oxford University Press.

Institute of Medicine. 1988a. The Future of Public Health. Washington, DC: National Academy Press.

Institute of Medicine. 1988b. Role of the Primary Care Physician in Occupational and Environmental Medicine. Washington, DC: National Academy Press.

Institute of Medicine. 1991. Addressing the Physician Shortage in Occupational and Environmental Medicine. Washington, DC: National Academy Press.

Institute of Medicine. 1993. Indoor Allergens: Assessing and Controlling Adverse Health Effects. Washington, DC: National Academy Press.

Institute of Medicine. 1995a. Environmental Medicine: Integrating a Missing Element into Medical Education. Washington, DC: National Academy Press.

Institute of Medicine. 1995b. Nursing, Health, and the Environment: Strengthening the Relationship to Improve the Public's Health. Washington, DC: National Academy Press.

Institute of Medicine. 1996. Interactions of Drugs, Biologics, and Chemicals in U.S. Military Forces. Washington, DC: National Academy Press.

Institute of Medicine. 1997. Toxicology and Environmental Health Information Resources. The Role of the National Library of Medicine. Washington, DC: National Academy Press.

Institute of Medicine. 1999. Cancer Research Among Minorities and the Medically Underserved. Washington, DC: National Academy Press.

Johnson, B., and S. Coulberson. 1993. Environmental epidemiologic issues and minority health. Annals of Epidemiology 3:175–180.

King, L. 1996. Presentation to the Institute of Medicine Committee on Environmental Justice, October 25, Chicago.

Kipen, H., and J. Craner. 1992. Sentinel pathophysiologic conditions: An adjunct to teaching occupational and environmental disease recognition and history taking. Environmental Research 59:93–100.

Kipen, H., D. Wartenberg, P. F. Scully, and M. Greenberg. 1991. Are non-whites at greater risk for occupational cancer? American Journal of Industrial Medicine 19:67–74.

Lee, C. 1992. Proceedings: The First National People of Color Environmental Leadership Summit. New York: United Church of Christ.

Lillie-Blanton, M., and S. C. Hoffman. 1995. Conducting and assessment of health needs and resources in a racial/ethnic minority community. Health Services Research 30:225–236.

Mannino, D. M., D. M. Homa, C. A. Pertowski, A. Ashizawa, L. L. Nixon, C. A. Johnson, L. B. Ball, E. Jack, and D. S. Kang. 1998. Surveillance for asthma— United States, 1960–1995. Morbidity and Mortality Weekly Report, CDC Surveillance Summary 47(SS-1):1–27.

Marshall, E. 1998. DNA studies challenge the meaning of race. Science 282:654–655.

Miller, B. A., L. N. Kolonel, L. Bernstein, J. L. Young Jr., G. M. Swanson, D. West, C. R. Key, J. M. Liff, C. S. Glover, and G. A. Alexander, eds. 1996. Racial/Ethnic Patterns of Cancer in the United States 1988–1992. Bethesda, MD: National Cancer Institute.

Montgomery, L.E., and O. Carter-Pokras. 1993. Health status by social class and/or minority status: Implications for environmental equity research. Toxicology and Industrial Health 9:729–773.

Motavalli, J. 1998. Toxic targets. E/The Environmental Magazine July–August:28–41.

National Center for Health Statistics. 1998a. Current Estimates from the National Health Interview Survey, 1995. Vital Health Statistics Series No. 10 (199). Hyattsville, MD: U.S. Public Health Service.

National Center for Health Statistics. 1998b. Health, United States, 1998 with Socioeconomic Status and Health Chartbook. Hyattsville, MD: U.S. Public Health Service.

National Environmental Justice Advisory Council. 1996. The Model Plan for Public Participation. Washington, DC: Environmental Protection Agency.

National Institute of Environmental Health Sciences. 1994. Symposium on Health Research and Needs to Ensure Environmental Justice Recommendations. February 10–12. Bethesda, MD: National Institute of Environmental Health Sciences.

National Library of Medicine. 1993. Improving Toxicology and Environmental Health Information Services. Report of the Board of Regents Long Range Planning Panel on Toxicology and Environmental Health. Bethesda, MD: National Library of Medicine.

National Research Council. 1984. Toxicity Testing: Strategies to Determine Needs and Priorities. Washington, DC: National Academy Press.

National Research Council. 1988. Complex Mixtures: Methods for In Vivo Toxicity Testing. Washington, DC: National Academy Press.

National Research Council. 1989a. Biologic Markers in Pulmonary Toxicology. Washington, DC: National Academy Press.

National Research Council. 1989b. Biologic Markers in Reproductive Toxicology. Washington, DC: National Academy Press.

National Research Council. 1989c. Improving Risk Communication. Washington, DC: National Academy Press.

National Research Council. 1991a. Animals as Sentinels of Environmental Health Hazards. Washington, DC: National Academy Press.

National Research Council. 1991b. Environmental Epidemiology, Vol. 1. Public Health and Hazardous Wastes. Washington, DC: National Academy Press.

National Research Council. 1992a. Biologic Markers in Immunotoxicology. Washington, DC: National Academy Press.

National Research Council. 1992b. Environmental Neurotoxicology. Washington, DC: National Academy Press.

National Research Council. 1993. Measuring Lead Exposure in Infants, Children, and Other Sensitive Populations. Washington, DC: National Academy Press.

National Research Council. 1996. Understanding Risk: Informing Decisions in a Democratic Society. Washington, DC: National Academy Press.

National Research Council. 1997. Environmental Epidemiology, Vol. 2. Use of the Gray Literature and Other Data in Environmental Epidemiology. Washington, DC: National Academy Press.

National Research Council. 1998. Research Priorities for Airborne Particulate Matter: I. Immediate Priorities and a Long-Range Research Portfolio. Washington, DC: National Academy Press.

Nez Perce Tribe. 1995. Our Environment, Our Health: How to Reduce Your Exposure to Toxic Substances in the Environment. Lapwai, Idaho: Nez Perce Tribe.

Ozonoff, D. 1995. Environmental medicine for all: Getting there from here. Lancet 346:860.

Parker, S. L., K. J. Davis, P. A. Wingo, L. A. G. Ries, and C. W. Heath. 1998. Cancer statistics by race and ethnicity. Cancer 48:31–48.

Presidential/Congressional Commission on Risk Assessment and Risk Management. 1997. Framework for Environmental Health Risk Management, Final Report, Vol. 1. Washington, DC: Presidential/Congressional Commission on Risk Assessment and Risk Management.

Rabin, S. A. 1994. A private view of health, surveillance and communities of color. Public Health Reports 109:42.

Redlich, C.A., W. S. Beckett, J. Sparer, K. W. Barwick, C. A. Riely, H. Miller, S. L. Sigal, S. L. Shalat, and M. R. Cullen. 1988. Liver disease associated with occupational exposure to the solvent dimethylformamide. Annals of Internal Medicine 108:680–686.

Rios, R., G. V. Poje, and R. Detels. 1993. Susceptibility to environmental pollutants among minorities. Toxicology and Industrial Health 9:797–820.

Robinson, F. 1996. Louisiana. Alsen, LA: North Baton Rouge Environmental Association.

Roe, D., W. Pease, K. Florini, and E. Silbergeld. 1997. Toxic Ignorance. New York: Environmental Defense Fund.

Rom, W. N., ed. 1997. Textbook of Environmental and Occupational Medicine, 3rd ed. Philadelphia: Lippincott-Raven.

Rosenstock, L., and M. R. Cullen, eds. 1994. Textbook of Clinical Occupational and Environmental Medicine. Philadelphia: W. B. Saunders.

Rutstein, D. D., R. J. Mullan, T. M. Frazier, W. E. Halperin, J. M. Melius, and J. P. Sestito. 1983. Sentinel health events (occupational): A basis for physician recognition and public health surveillance. American Journal of Public Health 73:1054–1062.

Sampson, R. J., W. R. Stephen, and F. Earls. 1997. Neighborhoods and violent crime: A multilevel study of collective efficacy. Science 277:918–924.

Schulte, P. A., and F. Perera. 1993. Molecular Epidemiology: Principles and Practice. New York: Academic Press.

Sexton, K., K. Olden, and B. L. Johnson. 1993. Environmental justice: The central role of research in establishing a credible scientific foundation for informed decision making. Toxicology and Industrial Health 9:685–727.

Stapleton, S. 1998. Asthma rates hit epidemic numbers; experts wonder why. American Medical News 41(18):4.

Szasz, A., and M. Meuser. 1997. Environmental inequalities: Literature review and proposals for new directions in research and theory. Current Sociology 45(3):99–120.

United Church of Christ Commission for Racial Justice. 1987. Toxic Wastes and Race in the United States: A National Report on the Racial and Socio-Economic Characteristics of Communities with Hazardous Waste Sites. New York: United Church of Christ.

U.S. Bureau of the Census. 1997. Results of the 1996 Race and Ethnic Targeted Test: Population Division Working Paper No. 18. Washington, DC: U.S. Bureau of the Census.

U.S. General Accounting Office. 1983. Siting of Hazardous Waste Landfills and Their Correlation with Racial and Socio-Economic Status of Surrounding Communities. Washington, DC: U.S. Government Printing Office.

U. S. Office of Mangement and Budget. 1978. Directive No. 15: Race and ethnic standards for federal statistics and administrative reporting. Statistical Policy Handbook. Washington, DC: U.S. Department of Commerce, Office of Federal Statistical Policy and Standards.

Vogel, G. 1997. Why the rise in asthma cases? Science 276:1645.

Wagener, D. K., D. R. Williams, and P. M. Wilson. 1993. Equity in environmental health: Data collection and interpretation issues. Toxicology and Industrial Health 9:775–795.

Waitzman, N. J., and K. R. Smith. 1998. Phantom of the area: Poverty-area residence and mortality in the United States. American Journal of Public Health 88:973–976.

Warren, R. C. 1993. The morbidity/mortality gap: What is the problem? Annals of Epidemiology 3:127–129.

Weiss, K.B., and D. K. Wagener. 1990. Changing patterns of asthma mortality: Identifying target populations at high risk. Journal of the American Medical Association 264:1683–1687.

Weitzman, M., S. Gortmaker, and A. Sobol. 1990. Racial, social, and environmental risks for childhood asthma. American Journal of Diseases of Children 144:1189–1194.

Wernette, D. R., and L. A. Nieves. 1993. Minorities and air pollution: A preliminary geodemographic analysis, 1991. Presented at the Socioeconomic Research Analysis Conference, Baltimore, Maryland. June 27–28; quoted in Sexton, K., H. Gong, J. C. Bailar, J. G. Ford, D. R. Gold, W. E. Lampert, and M. J. Utell. 1993. Air pollution health risks: Do race and class matter? Toxicology and Industrial Health 9:843–878.

Williams, D. R., R. Lavizzo-Mourey, and R. C. Warren. 1994. The concept of race and health status in America. Public Health Reports 109:26–41.

World Health Organization. 1986. Constitution. *In:* World Health Organization: Basic Documents. Geneva, Switzerland: World Health Organization.

Zahm, S. H., L. M. Pottern, D. R. Lewis, M. H. Ward, and D. W. White. 1994. Inclusion of women and minorities in occupational cancer epidemiologic research. Journal of Occupational Medicine 36:842–847.

Zimmerman, R. 1993. Social equity and environmental risk. Risk Analysis 13:649–666.

Zorpette, G. 1996. Hanford's nuclear wasteland. Scientific American 274:88–97.

Appendixes

A

Using Disease-Cluster and Small-Area Analyses to Study Environmental Justice

Daniel Wartenberg

This appendix explains, from a methodological point of view, why there are a paucity of epidemiologic or health-effects studies that assess environmental equity on a local, residential scale. It examines why the data requirements for traditional epidemiologic studies are rarely met and explains the strengths and limitations of screening for communities at greatest risk by using preepidemiologic studies (that is, studies with more limited data), such as disease-cluster and small-area analyses. Finally, it concludes with a set of research needs and a strategy for implementing health-effects studies to assess issues of environmental justice.

HEALTH-EFFECTS STUDIES

Barriers to Epidemiologic Studies

To understand why few studies have examined the health status of minority and economically disadvantaged populations living in contaminated environments, it is helpful to define what types of health-effects studies are possible and how scientists undertake such studies. There are two principal barriers to applying traditional epidemiologic methods to issues of environmental justice: data availability and sample size.

This Appendix is based on a paper by Daniel Wartenberg prepared for the committee.

Data Availability

To conduct an epidemiologic study of the effects of a local exposure source on a residential population, one needs to know the demographics of the population at risk, the extent of the exposure of concern, and information about other risk factors for disease, such as occupation, diet, and socioeconomic status. If the population is large enough and if a sufficient number of people are exposed to the disease agent, as defined by statistical criteria, then these data can be used in a traditional epidemiologic design (e.g., a cohort, case-control, or cross-sectional study). These epidemiologic studies help determine whether those people with higher exposures are or were more likely to develop disease when the effects of other risk factors for disease are removed or adjusted for. That is, one can compare residents with disease to those without disease in terms of exposure while making adjustments (either statistically or in the design) for other risk or lifestyle factors.

However, in most minority and low-income communities, such data are not readily available (Environmental Protection Agency, 1992; Sexton et al., 1993). As noted in an Environmental Protection Agency report on this issue, "Environmental and health data are not routinely collected and analyzed by income and race" (Environmental Protection Agency, 1992, p. 1). When such data are collected, they are not always available to researchers at the community or neighborhood level. For example, for some cancer incidence studies, states have been willing to provide municipal-level data and, in special circumstances, individual-level data. For studies in other states, data have been limited to the county level. Furthermore, data on lifestyle and behavior are not generally available except for regional data based on statistical sample surveys (e.g., the Centers for Disease Control and Prevention's Behavioral Risk Factor Surveillance System). Therefore, to undertake epidemiologically reliable studies of these communities, substantial data collection efforts would be required, but data collection is an extremely costly and resource-intensive enterprise.

Sample Size

In addition, many of the minority and low-income communities with environmental justice concerns are extremely small, in epidemiologic terms. Studies of small populations may not have adequate statistical power to detect a significant effect even if one exists. For example, Zimmerman reports that communities with at least 2,500 residents and at least one inactive hazardous-waste site on the National Priorities List (NPL) have a 1990 median population of less than 18,000 people (Zimmerman, 1993). To appreciate the limitations of studying a population this small or smaller, consider the following hypothetical calculations. For a cancer with an incidence rate of 1 in 30,000 people per year, the typical rate of childhood leukemia, one would expect to see fewer than one case per year in over half of the communities with NPL sites. Even if 5 years of health outcome data were available, one would expect to see fewer than three cases in each of these communities and about six cases in 10 years, a scant number of cases for reliable epidemiologic analysis.

However, it is not enough to know only about the people who developed adverse health effects. For most epidemiologic analyses, one also must have information about the people who did not develop disease so that explanations for the disease other than the one of interest, the environmental contamination, can be ruled out. To appreciate the concerns about statistical power (i.e., the ability to detect an effect statistically when one exists), consider the following hypothetical study. Assume that one can identify all childhood leukemia cases in one of these communities over a 5-year observation period and obtain data on other risk factors for the entire population, a rather big assumption. Then, assume (on the basis of typical U.S. data) that approximately 40 percent of a typical community is under 20 years of age. Then, assume that those exposed to the environmental contaminants were five times more likely to develop leukemia than those who were not exposed, a risk larger than that typically seen in environmental epidemiologic studies. With these data, an epidemiologic study would have a less than 50 percent chance of detecting a statistically significant effect. Studies with such a low likelihood of success are not attractive to researchers, are expensive, and are extremely difficult to fund. Small sample size is a major impediment to conducting meaningful studies.

Preepidemiology

As an alternative, before embarking on large, costly data collection efforts, it is sometimes possible to gain insight into local health risk by conducting preliminary studies with existing, albeit limited, data. This is called *preepidemiology* (Wartenberg and Greenberg, 1993). The methods used in these preepidemiologic assessments are typically called *disease-cluster* and *small-area analyses*. Some methods primarily use information about cases, not taking account of the characteristics of the population from which they were drawn. For example, if the locations or dates of incidence, or both, of a series of cases are known, it is possible to investigate whether these cases are closer together (or closer to a known exposure source) than would be expected. On the other hand, it might be possible to investigate whether the populations of three towns with borders within a mile of the local incinerator have higher rates of lung cancer than the lung cancer rate in the entire county in which they reside. Such studies can be used for the screening of populations in regions with a high incidence of a particular disease for further study and as an aid in the design of more rigorous studies. Because the data requirements for preepidemiologic studies are far more limited than those for traditional epidemiologic methods and because the data may be more subject to underreporting or inaccurate reporting, depending on the source, the results of the analyses are less reliable. Nonetheless, these data are still useful for screening and for designing additional studies.

Types of Preepidemiologic Methods

Cluster Data   In some preepidemiologic studies, data are analyzed without knowledge of the size of the population from which they came or whether there is a local source of risk. Such analyses simply summarize the characteristics of the patterns of cases. For other analyses one must determine the size of the population from which the cases have been identified. Sometimes, when a resident notices one or two cases of a rare disease in an area, he or she asks all neighbors and friends if they know of any other cases. When the observation is reported to health officials, the context (e.g., two cases on the resident's block) may overestimate the rate (i.e., these two cases may be the only ones in the entire town), making the disease rate look unnecessarily high. There also are issues of statistical stability of rate estimates if the population at risk is very small. Some more sophisticated disease-cluster methods use data about some of the people who did not develop disease, to adjust for other differences between the cases and those who did not develop disease (e.g., behavior and lifestyle). Some analyses compare average distances among those with disease to average distances among those without disease, whereas others use data on the proximity of people to a known exposure source. These are so-called focused studies (see below), which determine whether those with disease are closer to the source than those without disease.

Area Data   Another type of data used in preepidemiology is a set of disease rates reported for a set of geographic units, such as counties. These are called *area data*. Occasionally, such data are available at the geographic scale of the minor civil division (MCD), but most often the locational information is not sufficiently reliable at that level. This type of data can be analyzed in a variety of ways, such as to see if geographic units with higher rates of disease are nearer one another than expected or whether those closer to facilities emitting pollution have higher rates of disease than those farther away.

Limitations of Preepidemiology

One limitation of these summary data is that for exposures of very limited geographic or temporal extent, the majority of the MCD or county may be unexposed, diluting any possible excess cases of disease when data are reported for the unit as a whole. In addition, an exposure source is sometimes near a boundary of MCDs, so that several MCDs must be combined to capture the entire exposed population, even though only those living closest to the source are exposed, compounding the dilution problem even further. Another limitation of these data is that data on other risk factors most often are not available for individuals. Therefore, group summaries must be used, precluding adjustment for possible covariance of risk factors within a geographic unit, which leads to possible confounding (i.e., misattribution of cause [Greenland, 1989, 1992; Richardson et al., 1987; Rothman, 1990; Susser, 1994a,b]).

Although many of the preepidemiologic approaches make clever use of scant data, some scientists question the reliability and utility of such analyses (Neutra, 1990; Rothman, 1990). Because the limitations of each specific data set define its interpretableness and generalizability, these limitations must be made explicit both to researchers and to the community under study. This includes discussions with the community about the limitations of the methods before any data collection or data analyses are undertaken, lest false expectations be raised.

Appropriate Uses of Preepidemiology

On the other hand, preepidemiologic studies of case reports can be extremely useful as screening tools and for guidance in epidemiologic design. For example, in the 1850s, John Snow, a British physician, noticed higher death rates from cholera in one area of London than another. He hypothesized that this might be due to the source of the drinking water and its proximity to sewage disposal areas. By preventing access to the suspected contaminated water supply by removing the water pump handle, Snow was able to confirm his hypothesis (Snow, 1965).

Similarly, in Woburn, Massachusetts, in the early 1980s, residents reported excess cases of childhood leukemia that were confirmed by preepidemiologic analyses. A rigorous epidemiologic study further validated the residents' concerns and implicated chemically contaminated drinking water as the cause (Lagakos et al., 1986). The results of the latter study remain controversial, although a recent report supports the initial finding of cases of excess disease and reports a subsequent decline after appropriate allowance for latency.

In 1971, Herbst and coworkers reported on eight cases of adenocarcinoma of the vagina in women aged 15 to 22 in Massachusetts in which diethylstilbestrol was implicated as the disease-causing agent (Herbst et al., 1971). A year later, the observation of three cases of angiosarcoma of the liver among polyvinyl chloride production workers was used to implicate the vinyl chloride monomer as a cause of disease in a worker population (Creech and Johnson, 1974). Another workplace cluster of disease, male infertility in the pesticide industry, was used to identify the manufacture of dibromochloropropane as dangerous (Whorton et al., 1977). Several cases of phocomelia alerted experts to the problems of thalidomide. The identification of risk factors for human immunodeficiency virus transmission and AIDS also arose out of case reports from preepidemiologic studies (Centers for Disease Control, 1981). Finally, studies of soybeans and asthma attacks have identified soybeans as a new etiologic factor for the disease, and studies of Hodgkin's disease in young adults and mesothelioma in the small village of Karain, Turkey, have helped focus further epidemiologic studies that eventually led to a reduction in the number of cases of disease (Alexander, 1992). In short, although many question the utility of these preepidemiology studies, there are numerous examples of successes, that is, preliminary results confirmed by further rigorous epidemiologic studies.

STATISTICAL APPROACHES TO PREEPIDEMIOLOGY

Over 70 methods are used in disease-cluster and small-area analysis studies, and it is important to understand which type of method should be used in which situation, as well as the expected results and limitations of each method. I begin with a summary of the current practice of disease-cluster analysis, provide a set of questions to help investigators determine which analytic method is most appropriate for their study, briefly summarize the characteristics of the available methods, and conclude with a discussion of the statistical power of the methods.

Disease-Cluster Investigation Practices

To characterize the state of disease-cluster analyses, my colleagues and I have undertaken a review of published studies from 1960 to 1990. The goal was to characterize the problems studied, the approaches used, and the results obtained. The work is still under way, and the results presented here are preliminary.

A fairly wide variety of diseases were studied, although a few diseases were studied far more often than all the others. Of the 352 reported diseases, 207 (59 percent) were some type of cancer, 30 (8 percent) were some type of birth defect, and 14 (4 percent) were multiple sclerosis. (The area of infectious diseases is one that has largely been excluded in the preliminary review.) Of the studies of cancers, over one third were studies of leukemia and over one fifth were studies of Hodgkin's disease.

Approximately 74 different statistical analysis methods were used in the studies described in the set of 287 papers reporting statistical results. The two most commonly used methods, each used in over 20 percent of the reported studies, were the chi-square test (which compares the observed number of cases to that expected under an assumed Poisson distribution) and the Knox test for time-space interaction (Knox, 1963,1964a,b). Both of these methods were used more than twice as often as the next most frequently used method, the Ederer-Myers-Mantel (EMM) test for space-time clustering (Ederer et al., 1964). Only 12 of the 74 methods were each used in at least five separate studies. Few of the methods allow adjustment of confounding variables such as population density, ethnicity, age of residents, and so forth.

Of the categories of study types investigated, space-time clusters were assessed most often, followed by spatial clusters, seasonal periodicities, temporal clusters, and occupational clusters. A number of the reports were case series in which no statistical analyses were performed.

Overall, 71 percent of the reported results were statistically significant. This preponderance of positive findings has a few possible explanations. First, it may be that the reporting process overreports studies with positive findings. That is, although many neighborhoods evaluate their own disease frequencies (either explicitly or implicitly), those with excess disease frequencies are predominantly reported on and receive more rigorous consideration. Few communities, if any, report a deficit of cancers. So, even if the variation in cancer rates is random, with some neighbors showing higher than expected rates (on the basis of state or national averages) and some showing lower than expected rates, reporting of only the higher rates would produce a bias in the reports.

Second, it may be that studies with positive findings are more likely to be published. Many preliminary cluster investigations are conducted by health department officials and are never published. On the other hand, it may be that it is harder to have published a study with negative results than one with positive results. This, too, could create an overrepresentation of positive reports. However a substantial number (nearly one third) of studies with negative findings were published, suggesting that this is unlikely to fully explain the excess of studies with positive findings.

Choosing the Right Statistical Analysis Method

The statistical analysis methods most often used in preepidemiologic studies are the most simple to apply: the chi-square test and the Knox test. Neither method allows adjustment for confounding factors. Given the limited data available for most cluster studies in terms of both sample size and number of variables assessed, it seems reasonable that most investigators use methods that are not designed to emphasize subtle aspects of the data. Nonetheless, it is important that investigators choose the method that will be most sensitive for the detection of the pattern or process that they think underlies the concern, while appreciating the characteristics and limitations of the specific method chosen. Unfortunately, choosing among the more than 70 different methods that have been used in the published literature (Jacquez et al., 1996; Marshall, 1991) poses a challenge for researchers seeking to conduct an investigation. Below are six questions that may help researchers choose the most appropriate method (Waller and Jacquez, 1995; Wartenberg and Greenberg, 1993).

What Type of Clustering Is Being Investigated?

It is important that investigators choose the method that will be most sensitive for detection of the pattern or process that they think underlies the concern, while appreciating the characteristics and limitations of the specific method chosen.

Typical methods address time-, space-, space-time-, and exposure-based hypotheses. Assessments of patterns over time is the simplest, because data are distributed along one axis: time (Bailar et al., 1970; Ederer et al., 1964; Larsen et al., 1973; Naus, 1965; 1970; Tango, 1984). Patterns over time can be thought of as clusters (many events bunched together), trends (a gradual increase or decrease in the rate of events over time), or cycles (repeating patterns, such as high rates in certain seasons). For example, if one is concerned about a sudden increase in the incidence of asthma following the opening of a new industrial facility, one might want to look for a trend in asthma incidence over time, both before and after the opening. If one is concerned that events such as copycat suicides come in bunches, one might look for clusters of reported suicides using death certificates. If one believes that events come in cycles, such as asthma attacks during periods with high ozone levels in the summer, one might use a method that detects annual cycles.

Assessment over space is two–dimensional and has increased complexity compared with the assessment of patterns over time (Cuzick and Edwards, 1990; Diggle, 1991; Geary, 1954; Grimson et al., 1981; Moran, 1948, 1950; Ohno et al., 1979; Openshaw et al., 1988; Schulman et al., 1988; Tango, 1984; Whittemore et al., 1987). Spatial patterns may be clusters (regions with more events than other regions) or trends (gradual increases in the number of events across the study area). For example, one may suspect that there was a large spill of toxic material somewhere nearby but may not be sure where. Then, one might want to look for clusters of adverse health events that might signal the location (and effect) of the spill. On the other hand, one might be concerned about the confluence of emissions for a variety of pollution emitters, although one might not know the specific dispersion patterns of the mixed releases. For that situation, one might postulate both clusters of adverse health outcomes near the facilities, signifying hot spots, and a decreasing trend of adverse health effects as one moves away from the general area of the facilities, indicating the atmospheric processes of dilution, dispersion, and transport. In such cases, one would use spatial methods.

A specialized group of spatial methods assesses clustering in proximity to a particular source of hazard. They are called *focused methods* (Besag and Newell, 1991; Bithell and Stone, 1989; Lawson and Williams, 1993; Stone, 1988; Waller et al., 1994). Most often, these methods use distance as a surrogate for exposure and assess whether cases are closer to the source than expected. The advantage of these methods over other spatial methods is that they address a specific hypothesis of concern and, because of their specificity, have increased sensitivity. The disadvantage of these methods is that their specificity limits their ability to detect more general patterns of clustering.

Space-time assessments are three–dimensional (Abe, 1973; Barton et al., 1965; David and Barton, 1966; Klauber, 1975; Knox, 1964 a,b; Mantel, 1967; Pike and Smith, 1968, 1974; Pinkel and Nefzger, 1959; Pinkel et al., 1963; Symons et al., 1983; Williams, 1984). However, distances in space are not commensurate with distances in time, which further increases the complexity of the assessment problem. In general, space-time methods look for corresponding patterns of events that occur in both space and time. For example, finding that asthma cases occur when there are releases from an industrial facility and that those asthma cases tend to occur in specific geographic regions (say, mainly downwind of the facility) would be an example of space-time clustering. Space-time cluster methods are particularly popular because space-time clusters are thought to be more unusual than time-only or space-only clusters because they require simultaneous, correlated effects in two independent domains: space and time. Furthermore, the appearance of space-time clusters is unlikely to be attributable to other factors such as age, ethnicity, socioeconomic status, or even random fluctuations because they require effects in both space and time simultaneously.

What Is the Null Hypothesis?

The null hypothesis defines the pattern that one would expect to observe in a data set if there were no clustering. One might want to assume that cases are equally likely to occur at all locations. This is termed a *random, uniform risk distribution*. However, this assumption of uniform risk is overly simplistic. The total number of expected cases for any small spatial or temporal region is the sum of each individual's chance of being a case, or the individual's risk.

Since population density varies by block, by town, by county, and so forth, risk also varies. To the degree that one can model this variation in risk, the accuracy of the assessment can be improved. Beyond population density, it is known that demographic characteristics such as ethnicity, gender, and age also vary. Beyond demographics, it is known that behaviors and lifestyles vary. These factors—population density, demographics, behaviors, and lifestyles—are called *confounding variables* because they might be alternative explanations for the observed cases of disease. To the degree that one can capture these sources of variation, one can further improve the accuracy.

For temporal cluster methods, many methods assume the random, uniform risk distribution as the null hypothesis (Grimson and Rose, 1991; Grimson et al., 1992; Larsen et al., 1973; Naus, 1965; Tango, 1984; Wallenstein, 1980). That is, in each equal interval of time one expects to observe the same number of cases, with that number being the total number of observed cases divided by the number of equal time intervals. A modification of this method that allows investigators to accommodate confounding variables uses unequal time intervals in which the size of each time interval is set according to the population size (or some other function of confounding variables), still keeping the number of expected cases equal for each (unequal) time interval (Weinstock, 1981).

Most spatial methods similarly assume the random, uniform risk distribution as the null hypothesis (Cliff and Ord, 1981; Geary, 1954; Grimson and Rose, 1991; Moran, 1948, 1950). Some methods enable researchers to modify the size of the spatial interval to accommodate confounding variables such as population size (Hjalmars et al., 1996; Kulldorff, 1997; Openshaw et al., 1988; Turnbull et al., 1990). Some methods modify the distance between geographic units to reflect the underlying population density by a statistical (Whittemore et al., 1987) or graphical (Schulman et al., 1988) means. One method weights comparisons between nearby units by their relative population densities (Oden,

1995). Others enable investigators to choose a set of subjects without disease (i.e., controls) and then compare the distribution of the cases to that of the controls (Cuzick and Edwards, 1990; Diggle, 1991). One method characterizes the demographics of an entire neighborhood population, which was then used to define the risk of disease for each individual (Day et al., 1988).

Some space-time clustering methods determine whether cases are close in space and time simultaneously, close only in space, close only in time, or close in neither, assuming as the null hypothesis that closeness in space and closeness in time are unrelated (Barton et al., 1965; Klauber, 1975; Knox, 1963, 1964; Mantel, 1967; Marshall, 1991; Pinkel and Nefzger, 1959; Symons et al., 1983; Williams, 1984). Pinkel and coworkers (1963) refined this approach by defining the null distribution of time differences between cases as that observed in cases that were spatially far apart and comparing the distribution of time differences among cases spatially close to it. Mantel (1967) extended this approach by using linear regression to compare the time differences between all pairs of cases to their spatial distances, and Jacquez (1995) modified it to compare nearest neighbors in space to those in time. Finally, some methods sample a set of individuals from the same study area but without disease (i.e., controls) and use their distribution as the null hypothesis against which the distribution of cases is compared (Lyon et al., 1981; Pike and Smith, 1974). The latter approach is most able to capture unknown confounding variables.

What Is the Alternative Hypothesis?

The alternative hypothesis defines the pattern that one wishes to infer if one rejects the null hypothesis. Statistical methods are not equally powerful against all alternative hypotheses. Typically, certain methods are more sensitive at detecting a particular type of pattern, such as a single, large cluster of cases, whereas others are more sensitive at detecting another type of pattern, such as many small clusters of cases distributed throughout the study area. Other possible alternative hypotheses are that the cases follow the shape of an exposure source, such as high-voltage electric power transmission lines, polluted rivers, or the air dispersion pattern downwind of a single stack. One also could imagine a trend in which the risk of disease falls off gradually with increasing distance from a certain location or the source of exposure. In choosing a method, it is useful to review the results of tests of statistical power (see below) to determine which method is most sensitive for the particular type of pattern being investigated.

What Type of Summary Statistic Is Used?

Two main statistical approaches are used by cluster methods. In one, distances between all pairs of events (case-case, case-control, case-exposure, etc.) are calculated and summarized. These may be summarized as the number of case pairs closer than a certain separation distance, such as the mean distance between cases or such as the mean distance of each case to the nearest other case. On the other hand, the whole distribution of distances may be used. If controls are used, then a summary of the case distances may be compared to a summary of the control distances or even to case-control distances.

In the other approach, called the *cell occupancy approach*, the study area (i.e., time, space, or space-time) is divided into a set of cells, and each is assigned an expected number of cases on the basis of the overall disease rate, possibly adjusting for population density or other risk factors. Then, the observed number of cases in each cell is compared to the expected number in each cell.

Which Method Is Most Appropriate for the Cluster Under Investigation?

Two main characteristics are used to evaluate preepidemiologic methods: bias and sensitivity. When one is planning to undertake a preepidemiologic investigation, it is important to evaluate these characteristics so that one appreciates the strengths and limitations of the method. When working with communities, it is at least as important to discuss these characteristics with the community representatives, so that they have an understanding of what the method can and cannot do. If the residents do not appreciate the limitations at the outset and the study gives a negative result, the community representatives may believe that the scientists are hiding something and that they purposefully used a method that could not find a cluster. However, if they understand the limitations at the outset, they can either accept or reject the methodological approach independent of the results, a far more objective and satisfying evaluation. On the other hand, researchers could work with residents to select a method that both is appropriate scientifically and accommodates residents' preferences. These characteristics are now considered.

Bias, the finding of a false effect or the obscuring of a real effect, is a critical factor of concern. For example, if investigators conducted a study of lung cancer incidence around an industrial facility, they might find a cluster of disease near the facility. However, it might be that data on smoking were not available, and it might be that a greater proportion of the population that lived close to the facility than those that live far away were smokers (due to clustering of people of similar ethnicity or socioeconomic status). If the investigators had been able to adjust for smoking, the cluster may have been fully explained, exonerating the industrial facility. Therefore, by not adjusting the data for smoking patterns, the investigators drew the wrong conclusion. Unfortunately, there is no easy way to test for this type of bias unless data on all possible factors that could cause the disease are available. However, when reliable data on some possible risk factors are available, it is better to use methods that adjust for these data than methods that do not.

Sensitivity (i.e., statistical power) is the ability of a method to detect an effect when it really exists. Sometimes, a study with very few cases is conducted. The method may not show statistically significant clustering even though a real cluster exists because random variation could produce a similar effect. Ideally, methods should be extremely sensitive. To evaluate the sensitivity of a method, investigators conduct computer simulations. In a set of simulations, investigators construct a large number of hypothetical data sets that have a number of cases that cluster. Then, the investigators add a number of cases that do not cluster but that are distributed randomly. Typically, one constructs several thousand of these hypothetical data sets. Next, the method being evaluated is applied to all of these hypothetical data sets, and the percentage of times that the method detects the cluster is reported. The percentages are then subdivided so that one can determine how the sensitivity of the method decreases as the proportion of random cases increases. Studies that have compared the sensitivities of different methods are described below.

In conducting studies to evaluate the sensitivity or statistical power of a method, the investigator sets three important parameters: the number of events, the relative risk, and the alternative hypothesis. Most often, in preepidemiologic investigations, there are a few to a few dozen cases. These are small numbers for any epidemiologic study. Most simulation studies evaluate 25 to 250 cases.

The relative risk is the typical measure of the severity of the hazard. It tells how much more likely a person is to get the disease if the person is exposed to the hazard. Risks of known environmental hazards usually range from about 1 to 5, with unusually strong hazards being as large as 10. For preepidemiologic methods, in which exposure most often is not well characterized, most investigators would like to have sensitivity to detect risks of at least 3 or higher. Some investigators postulate more complex models of epidemicity, particularly for infectious diseases, which preclude translation into simple relative risks. In short, the relative risks used in simulation studies vary widely.

The alternative hypothesis, as noted above, is the specific pattern that the investigator creates in the hypothetical data sets for the simulations (e.g., a single cluster, multiple clusters, trend, results of models of pollutant dispersion, and results of models of specific disease spread). Some investigators use simplistic models (e.g., single cluster hot spots or trends). These are easy to explain but are unrealistic. Other investigators use complicated models of contagions or the disease process. Still others use data sets that are modified from actual studies. These may capture subtle characteristics of data that may be hard to identify and that are difficult to explain or describe.

The results of statistical power studies provide two types of information. First, for a specific data pattern (e.g., a single cluster or a trend), they allow investigators to compare the sensitivities of methods. Typically, one method is more sensitive for one type of pattern and another method is more sensitive for another type of pattern. Second, they allow investigators to estimate the minimum number of cases needed to detect an effect, given a rather strict set of assumptions. Unfortunately, different investigators often use different sets of assumptions, which result in the need for different minimum numbers of cases.

For example, in studies of time patterns, Naus (1966) found that a method that assesses the interval with the greatest number of events (the scan method [Naus, 1965]) was more sensitive than a method that compares the number of observed cases to that expected for each time interval in the study EMM [Ederer et al., 1964]) for data sets with very few cases, but the reverse was true for data sets with many cases. Sometimes, results of these studies are highly dependent on a set of assumptions. Wallenstein and colleagues (1993), in a complementary investigation of the scan method (Naus, 1965), reported that for a hypothetical data set with a relative risk of 4, one needed only 10 cases for a sensitivity of 80 percent, whereas Sahu and colleagues (1993), using the same method with slightly different assumptions of what the pattern looked like, reported that with a relative risk of 4 one needed 50 cases for a sensitivity of 80 percent. Apparently, the scan method is quite sensitive to the definition of the cluster.

For space-time methods, several investigators have compared two or three methods using specific models of epidemicity (typically infectious diseases) or specific data sets (typically chronic diseases) (Alexander, 1991; Bithell, 1992; Cartwright et al., 1990; Chen et al., 1984; Cuzick and Edwards, 1990; McAuliffe and Afifi, 1984; Raubertas, 1988; Shaw et al., 1988). Although these results highlight certain strengths of the most powerful methods, it is difficult to describe the results so that they are understandable or apply them to situations involving noncontagious diseases.

To investigate both spatial and space-time clustering, Wartenberg and Greenberg (1990a,b, 1992) compared a method using the distance between all cases (Mantel's space-time regression method [Mantel, 1967]), a method comparing the number of observed and expected cases in each space-time interval (the EMM method [Ederer et al., 1964, and two indices of spatial autocorrelation (Moran's I [Moran, 1948, 1950] and Geary's c [Geary, 1954]). They found that the method comparing observed and expected cases was more sensitive to trends in data than the other methods and that the distance-based method was more sensitive to single clusters. However, even with a relative risk of 5, to achieve adequate power (>80 percent) one needs sample sizes of at least 25 to 40 cases. In short, all methods were relatively insensitive.

Waller and Lawson (1995) studied the statistical power of focused cluster tests using hypothetical data sets with 51, 150, and 300 people with disease. A type of trend test (the local score tests [Lawson and Williams, 1993; Waller et al., 1994]) did best, achieving 80 percent power with risks of over 2 for the 300-case scenario for more concentrated clusters. However, the methods did not do as well with a risk of 4 for the 51-case situation. For a more diffuse cluster with the same number of total cases, the sensitivity is much greater.

Walter (1992a,b) investigated the possible bias arising from differences in population density (and, hence, the statistical stability of rate estimates) in area data and found a substantial effect. Oden and colleagues (1996), investigating Oden's population adjustment method for spatial autocorrelation (Ipop [Oden, 1995]), found the method to be more powerful than both unadjusted methods (Moran, 1948, 1950) and methods that use individual case data rather than area summaries (Cuzick and Edwards, 1990; Grimson and Rose, 1991; Grimson et al., 1981). The latter result is somewhat surprising and warrants further investigation. Even so, for simulations by this method with 30 cases, the power was moderate at best, generally under 50 percent.

In summary, cluster detection methods work best if there are at least 30 cases, and preferably at least 50 cases. Even so, the ability to detect an excess incidence of disease is dependent both on the choice of method and on the specific nature of the excess disease incidence. Some methods are powerful for one type of pattern (e.g., a single local excess disease incidence or cluster) and not powerful for another type of pattern (e.g., a more general increasing trend in risk or disease incidence with proximity to the source), and vice versa. As with most statistical tests, as one increases the specificity of the alternative hypothesis under investigation, such as with the focused tests, one increases the power for the detection of patterns consistent with that hypothesis while decreasing the power for the detection of patterns that are not consistent with that alternative hypothesis but that are still aberrant.

Two limitations of this whole body of literature are the lack of consistency and comparability of the simulation methods and the narrow focus of most simulation studies. Investigators have used simulated disease patterns, models of disease spread, and examples based on real data sets as alternative hypotheses for testing. Within each of these approaches, small differences in methods make comparisons difficult, and comparisons between these approaches are not direct. Furthermore, investigators usually evaluate only a few methods, and when they do so, most often they use methods designed to be sensitive to the same aspect of the pattern. Typically, the studies do not compare spatial methods to space-time methods to focused methods and do not use a wide array of data that simulate patterns for which each is likely to be the most powerful. By so doing, one could begin to evaluate the limitations of applying the wrong method to a given situation (i.e., compare the costs of specificity across methods). Although such an exercise may not be that interesting theoretically, it would be of substantial value to the practitioner.

How Should the Statistical Significance of an Analysis Be Interpreted?

Statistical interpretations of disease-cluster and small-area analyses are extremely controversial. As demonstrated by participants at the 1989 conference on disease clusters sponsored by the Centers for Disease Control, many investigators are skeptical of the utility and statistical reliability of cluster studies (Neutra, 1990; Rothman, 1990). Some investigators are concerned that cluster studies most often result from community awareness of high disease rates that occur as a result of temporary, random fluctuations. These observations are based on informal evaluations conducted in each community as neighbors talk to one another about their lives. Therefore, those that garner scientists' attention have been screened out of thousands and thousands of potential study areas. From a statistical inference point of view, one must make an adjustment in the calculation of the p value to adjust for the screening of thousands of communities to limit the multiple-comparisons problem and the likely false-positive results (Armon et al., 1991; Neutra, 1990). Others point out that the specificities of the methods may impede detection of clusters not exactly fitting the assumptions of the specific method being used. They recommend against ruling out the existence of a cluster even if the p value exceeds the nominal level (Grimson et al., 1992). In short, the problems of false-positive and false-negative results are far from resolved or even well understood in the community setting (Wartenberg, 1994).

In developing an interpretation of the statistical importance of a disease-cluster or small-area study, one must consider the goal of the evaluation and the use to which the result will be put. If used for screening (Wartenberg and Greenberg, 1993) or even for the identification of new etiologies (Rothman, 1990), the use of p values can be thought of as a way of ranking the severity of the clustering rather than traditional use for statistical interpretation. Then, by using this ranking in conjunction with other information about the observation (e.g., sample size and plausibility in terms of either known exposures or biologic models), situations can be targeted for more rigorous investigation. It is hoped that this further investigation, although expensive, would resolve the concern.

Another instance that raises the issue of p values is the follow-up of an assessment of the presence of an excess incidence of disease. Often, once an excess incidence is detected, prospective observation is instituted, and after several years of additional data collection, the new data are assessed. This is a response to community concerns that provides useful data for assessment. In fact, this is what happened in a study of childhood leukemia in Woburn, Massachusetts (Lagakos et al., 1986), in which an excess incidence of childhood leukemia appears to have persisted, even though the putative exposure, a contaminated drinking water well, was closed. (More epidemiologic research to understand this situation is under way.) In such situations, should one use a more liberal statistical criterion to evaluate the newly collected, prospective data since this community had already shown an excess incidence of childhood leukemia? Certainly, it should not be held to the same statistical standard as a new investigation in a situation with no history of an excess incidence. Should the tests be one tailed, since the a priori hypothesis is that there is an excess? Should the statistical criteria for assessments of disease causation hypotheses differ from those designed to guide public health policy? Should the confirmation of occurrence of a statistically significant excess incidence of disease be required in situations of known exposures before preventive action is undertaken? These questions pose difficult challenges.

In the context of environmental justice issues, the interpretations are also complex. These are situations in which a priori concerns based on known exposures may have resulted in a high incidence of disease. However, there are thousands of such situations throughout the United States. It is not clear how one should address these same issues of statistical significance testing. On the one hand, since one has an a priori concern, one might argue for doing one-tailed significance testing. If one is considering public health policy and disease prevention rather than etiologic research, one also might consider increasing the acceptable p value to diminish false-negative results. On the other hand, one might be concerned about false-positive results because of the large number of communities from which data for the study have been drawn. Then one would lower the p value to adjust for the multiple tests that have been carried out, albeit informally.

In short, one should not preclude the use of preepidemiologic studies of communities with environmental justice concerns on the basis of the statistical sensitivity of the methods, as described above. The issues are too complex and too poorly understood to dismiss this potentially useful methodology. The only criterion for a successful investigation should not be the identification of a new etiology (Neutra, 1990) or even the identification of an excess incidence of disease. Rather, helping to target more in-depth investigations and even just to put community concerns in focus can be an extremely useful process. Such exercises sometimes identify particular exposure pathways that can be addressed whether or not an excess incidence of disease is confirmed. Despite their limitations, preepidemiologic methods are useful for screening and for focusing hypotheses. The examples of preepidemiologic studies described above provide further justification for their successful use with more traditional standards, although many have argued that, in epidemiology, one should not be overly concerned with p values in the context of traditional inference (Rothman, 1990b; Savitz and Olshan, 1995; Thomas et al., 1985).

RESEARCH NEEDS

This summary of methods for the investigation of environmental justice issues has highlighted a number of limitations of these methods and needs for improvement. First and foremost is the need for better access to existing data. As noted above, states have different rules regarding access to individual data records with personal identifiers. Such data, when available, provide the best resource for researchers investigating local problems. These data need to be made available regardless of the researchers' professional affiliation or other political and bureaucratic issues, provided that the request is filed by a bona fide researcher and that adequate steps will be taken by the researcher to protect confidentiality.

Second, there is a need to develop better and more comprehensive data resources. Typically, small communities have data only on births and deaths, and often, these data are available only at the scale of the municipality or the county. Researchers need data for each individual with residential addresses or specific residence locations for analysis of disease incidence patterns and assessment of proximity to possible sources of exposure. With the advent of geographic information systems, such analyses are becoming increasingly easy to conduct (Guthe et al., 1992; Rushton et al., 1995, 1996; Wartenberg, D. 1992; Wartenberg et al., 1993; Wartenberg, D. 1994), although in most places such data are not available or of insufficient accuracy for meaningful analysis. In addition, it would be useful to have other data, such as cancer incidence and birth defects data, and data on less severe outcomes that might be affected by air pollution, such as asthma incidence and the results of pulmonary function tests. Finally, data on confounding variables, such as behavioral and lifestyle characteristics, would facilitate more rigorous evaluations. There exist several preepidemiologic methods that allow for adjustment by confounders, if such data are available. These adjustments may reveal hidden associations or explain associations that had erroneously been attributed to exposure to environmental contaminants. If such data are not made available, substantial resources will be needed to undertake investigations.

Third, there is a need for a more systematic evaluation of the statistical properties of the preepidemiologic methods. Investigators need to know what method to use when and what they can and cannot detect. The statistical power studies conducted to date do not provide that type of systematic evaluation. This evaluation could be facilitated by developing a protocol for such comparisons, including the specific data sets, hypothetical and real, that should be used and the test results that should be reported. Then, data or computer programs should be made available to the scientific community for more comprehensive testing of existing methods and for development and testing of new methods. Finally, by compiling the results, one may begin to understand how to use these tools and what interpretations to draw from the results.

Fourth, one must consider the trade-offs involved in the interpretation of statistical significance in preepidemiologic studies. Researchers have argued in different contexts that traditional significance testing is both too liberal and too conservative. There is a need to look at the type of problems for which these methods are used (e.g., identification of new etiologies, identification of specific exposures or excess incidence of disease, replication of observed historical excess incidences of disease, and development of public policy) and to develop guidelines for interpretation for each use. Simply applying the rule that statistical significance is a p value of <0.05, as is currently practiced, does not adequately address the disparity of needs and the variation in the severity of the consequences of false-positives and false-negative results. In epidemiology, results identifying a new etiology become credible only after substantial replication, regardless of the p value.

Failure to identify an exposure that is causing an excess incidence of disease (a false-negative result) will likely lead to additional cases of disease, possibly a high cost to society, while false confirmation of a hazard (a false-positive result) will lead to overprotection, possibly a relatively low cost. Public health policy typically dictates a greater willingness to accept false-positive alarms than false-negative missed diagnoses. However, cluster investigators do not consider these trade-offs.

Fifth, there is a need for the development and testing of methods for combining the results from studies performed at disparate locations (Neutra, 1990). In the past two decades, there has been an explosion in the use of meta-analysis, a method for combining the results of published studies. Although many criticize this methodology by pointing out its limitations, in appropriate contexts the approach has been very powerful. For environmental justice issues, at the current time, meta-analysis is not the appropriate tool. Too few studies have been undertaken, let alone published. However, if one could identify exposures with similar characteristics in several communities, each of which is too small individually to have adequate statistical power, the joint analyses of these communities' data might prove insightful. Such approaches have been tried in other situations in which the exposures are similar (Cardis et al., 1995; Geschwind et al., 1992; Marshall et al., 1995), but the methods and their limitations need to be examined in the context of using preepidemiologic methods and studying issues related to environmental justice. These include small sample sizes, comparability of communities, possible confounding, similarity of exposures, and so forth.

In conclusion, when conducting studies of adverse health effects in minority and economically disadvantaged communities, the following considerations apply:

• When concerns are raised, involve the community in all discussions and explain the limitations of the available methods and the possible outcomes of study.

  • Conduct a preepidemiologic assessment.
  • Use the right method for the job.
  • Be careful in evaluating the results.
  • Conduct an epidemiologic study.

The results of preepidemiologic studies should be evaluated in the context of the answers to the set of questions listed earlier. For example, if a method used is sensitive to trends but not clusters and negative results are obtained, one should not rule out the existence of clusters.

Similarly, results should be evaluated in the context of additional information. For example, if the observed health effect has not previously been reported for the hypothesized exposure, one should be far less confident of an association. If an excess incidence of disease is not observed where and when it was expected, one must question both the hypothesized association between this exposure and the excess incidence of disease and the adequacy of the exposure information. The association may be real, but it may be that researchers misunderstood how people were exposed.

The results also should be evaluated in the context of future risk. For example, if it is likely that people in the community are still being exposed to the suspected agent, one should consider taking remedial action more aggressively than one would if exposure were no longer occurring.

On the basis of these and related issues, the community, scientists, and policymakers must decide whether sufficient information has been gathered, whether exposure can and should be reduced, and whether further study is needed. Further study can mean working with officials to get access to existing databases that were not made available earlier or collecting new data and performing a new analysis.

If further study is needed, a traditional epidemiologic study should be conducted. The design of that study should be determined by the specifics of the situation. For example, if only data at the county level were available previously, the most useful study might try to get data on a more local level. This may require researchers to interview all current and former residents to get disease incidence and risk factor information. On the other hand, it may require working with local officials to break down regional data to the individual level. Alternatively, if data had previously been made available on a local level and there were enough cases for sufficient statistical power, effort might best be spent measuring or quantifying exposures.

An additional option is to implement a surveillance or sentinel investigation program. A surveillance program could be used routinely to develop the data necessary for investigation of health concerns, including ones not currently reported, and then to perform preepidemiologic assessments on a regular basis. This would provide a baseline from which one or several communities could monitor their health status. Appropriate caveats would have to be provided, so that yearly or local fluctuations were not interpreted as meaningful unless they were based on a sufficient number of cases.

In addition, one could develop a sentinel reporting system. This system would record occurrences of easily identifiable health events that may be indicators of environmental exposures but not contained in current reporting systems (Rothwell et al., 1991). However, to be useful, one would have to do more research to determine whether appropriate and useful sentinels can be identified.

BIBLIOGRAPHY

Abe, O. 1973. A note on the methodology of Knox's test of "time and space interaction." Biometrics 29:68–77.

Alexander, F. E. 1991. Investigations of localised spatial clustering and extra-poisson variation. *In*: The Geographical Epidemiology of Childhood Leukemia and Non-Hodgkin's Lymphomas in Great Britain 1966–1983. G. Draper, ed. London: Her Majesty's Stationery Office.

Alexander, F. E. 1992. Space-time clustering of childhood acute lymphoblastic leukaemia: Indirect evidence for a transmissible agent. British Journal of Cancer 65:589–592.

Armon, C., J. Daube, P. O'Brien, L. Kurland, and D. Mulder. 1991. When is an apparent excess of neurologic cases epidemiologically significant? Neurology 41:1713–1718.

Bailar, J. C., III, H. Eisenberg, and N. Mantel. 1970. Time between pairs of leukemia cases. Cancer 25:1301–1303.

Barton, D. E., F. N. David, and M. Herrington. 1965. A criterion for testing contagion in time and space. Annals of Human Genetics 29:97–103.

Besag, J., and J. Newell. 1991. The detection of clusters in rare diseases. Journal of the Royal Statistical Society Series A 154(part 1):143–155.

Bithell, J. F. 1992. Statistical methods for analysing point-source exposures. In: Geographical and Environmental Epidemiology: Methods for Small-Area Studies. P. Elliott, J. Cuzick, D. English, and R. Stern, eds. New York: Oxford University Press.

Bithell, J. F., and R. A. Stone. 1989. On statistical methods for analysing the geographical distribution of cancer cases near nuclear installations. Journal of Epidemiology and Community Health 43:79–85.

Cardis, E., E. S. Gilbert, L. Carpenter, G. Howe, I. Kato, B. K. Armstrong, et al. 1995. Effects of low doses and low dose rates of external ionizing radiation: Cancer mortality among nuclear industry workers in three countries. Radiation Research 142:117–132.

Cartwright, R. A., F. E. Alexander, P. A. McKinney, and T. J. Ricketts. 1990. Leukemia and Lymphoma: An Atlas of Distribution Within Areas of England and Wales 1984–1988. London: Leukemia Research Fund.

Centers for Disease Control. 1981. Pneumocystis Pneumonia—Los Angeles. Morbidity and Mortality Weekly Report 30:250–252

Chen, R., N. Mantel, and M. A. Klingberg. 1984. A study of three techniques for time-space clustering in Hodgkin's disease. Statistics in Medicine 3:173–184.

Cliff, A. D., and J. K. Ord. 1981. Spatial Processes: Models and Applications. London: Pion.

Creech, J. L., Jr., and M. N. Johnson. 1974. Angiosarcoma of the liver in the manufacture of polyvinyl chloride. Journal of Occupational Medicine 16:150–151.

Cuzick, J., and R. Edwards. 1990. Spatial clustering for inhomogeneous populations. Journal of the Royal Statistical Society, Series B 52:73–104.

David, F. N., and D. E. Barton. 1966. Two space-time interaction tests for epidemicity. British Journal of Social Medicine 20:44–48.

Day, R., J. H. Ware, D. Wartenberg, and M. Zelen. 1988. An investigation of a reported cancer cluster in Randolph, Massachusetts. Journal of Clinical Epidemiology 42:137–150.

Diggle, P. J. 1991. A point process modelling approach to raised incidence of a rare phenomenon in the vicinity of a pre-specified point. Journal of the Royal Statistical Society Series A 153:349–362.

Ederer, F., M. H. Myers, and N. Mantel. 1964. A statistical problem in space and time: Do leukemia cases come in clusters? Biometrics 20:626–638.

Environmental Protection Agency. 1992. Environmental equity: Reducing risk for all communities. In: Workgroup Report to the Administrator, Vol. 1. R. M. Wolcott and W. A. Banks, eds. Report EPA230-R-92-008. Washington, DC: Environmental Protection Agency.

Geary, R. C. 1954. The contiguity ratio and statistical mapping. The Incorporated Statistician 5:115–145.

Geschwind, S. A., J. A. J. Stolwijk, M. Bracken, E. Fitzgerald, A. Stark, C. Olsen, et al. 1992. Risk of congenital malformations associated with proximity to hazardous waste sites. American Journal of Epidemiology 135:1197–1207.

Greenland, S. 1989. Modeling and variable selection in epidemiologic analysis. American Journal of Public Health 79:340–349.

Greenland, S. 1992. Divergent biases in ecologic and individual-level studies. Statistics in Medicine 11:1209–1223.

Grimson, R. C., and R. D. Rose. 1991. A versatile test for clustering and a proximity analysis of neurons. Methods of Information in Medicine 30:299–303.

Grimson, R. C., K. C. Wang, and P. W. C. Johnson. 1981. Searching for hierarchical clusters of disease: Spatial patterns of sudden infant death syndrome. Social Science and Medicine 15D:287–293.

Grimson, R. C., T. E. Aldrich, and J. W. Drane. 1992. Clustering in sparse data and an analysis of rhabdomyosarcoma incidence. Statistics in Medicine 11:761–768.

Guthe, W. G., R. K. Tucker, E. A. Murphy, R. England, E. Stevenson, and J. C. Luckhardt. 1992. Reassessment of lead exposure in New Jersey using GIS technology. Environmental Research 59:318–325.

Herbst, A., H. Ulfelder, and D. Poskanzer. 1971. Adenocacinoma of the vagina. Association with maternal stilbestrol therapy with tumor appearance in young women. New England Journal of Medicine 284:878–881.

Hjalmars, U., M. Kulldorff, G. Gustafsson, and N. Nagarwalla. 1996. Childhood leukemia in Sweden: Using GIS and a spatial scan statistic for cluster detection. Statistics in Medicine 15:707–716.

Jacquez, G. M. 1995. The map comparison problem: Tests for the overlap of geographic boundaries. Statistics in Medicine 14:2343.

Jacquez, G. M., L. A. Waller, R. Grimson, and D. Wartenberg. 1996. The analysis of disease clusters. Part I. State of the art. Infection Control and Hospital Epidemiology 17:317–327.

Klauber, M. R. 1975. Space-time clustering analysis: A prospectus. Philadelphia: Society of Industrial and Applied Mathematics.

Knox, G. 1963. Detection of low intensity epidemicity: Application to cleft lip and palate. British Journal of Preventive and Social Medicine 17:121–127.

Knox, G. 1964a. The detection of space-time interaction. Applied Statistics 13:25–29.

Knox, G. 1964b. Epidemiology of childhood leukemia in Northumberland and Durham. British Journal of Preventive and Social Medicine 18:17–24.

Kulldorff, M. 1997. A spatial scan statistic. Communications in Statistics: Theory and Methods 26:1481–1496.

Lagakos, S., B. Wessen, and M. Zelen. 1986. An analysis of contaminated well water and health effects in Woburn, Massachusetts. Journal of the American Statistical Society 81:583–596.

Larsen, R. J., C. L. Holmes, and C. W. Heath. 1973. A statistical test for measuring unimodal clustering: A description of the test and of its application to cases of acute leukemia in metropolitan Atlanta, Georgia. Biometrics 29:301–309.

Lawson, A. B., and F. Williams. 1993. Applications of extraction mapping in environmental epidemiology. Statistics in Medicine 12:1249–1258.

Lyon, J. L., M. R. Klauber, W. Graff, and G. Chiu. 1981. Cancer clustering around point sources of pollution: Assessment by case-control methodology. Environmental Research 25:29–34.

Mantel, N. 1967. The detection of disease clustering and a generalized regression approach. Cancer Research 27:209–220.

Marshall, E. G., L. J. Gensburg, N. S. Geary, D. A. Deres, and M. R. Cayo. 1995. Analytic Study to Evaluate Associations Between Hazardous Waste Sites and Birth Defects. Atlanta: Agency for Toxic Substances and Disease Registry.

Marshall, R. J. 1991. A review of methods for the statistical analysis of spatial patterns of disease. Journal of the Royal Statistical Society Series A 154:421–441.

McAuliffe, T. L., and A. A. Afifi. 1984. Comparison of nearest neighbor and other approaches to the detection of space-time clustering. Computational Statistics and Data Analysis 2:125–142.

Moran, P. A. P. 1948. The interpretation of statistical maps. Journal of the Royal Statistical Society Series B 10:243–251.

Moran, P. A. P. 1950. Notes on continuous stochastic phenomena. Biometrika 37:17–23.

Naus, J. I. 1965. The distribution of the size of the maximum cluster of points on a line. Journal of the American Statistical Association 60:532–538.

Naus, J. I. 1966. A power comparison of two tests of non-random clustering. Technometrics 8:493–517.

Neutra, R. 1990. Counterpoint from a cluster buster. American Journal of Epidemiology 132:1–8.

Oden, N. L. 1995. Adjusting Moran's I for population density. Statistics in Medicine 14:17–26.

Oden, N., G. M. Jacquez, and R. Grimson. 1996. Realistic power simulations compare point- and area-based disease cluster tests. Statistics in Medicine 15:783–806.

Ohno, Y., K. Aoki, and N. Aoki. 1979. A test of significance for geographic clusters of disease. International Journal of Epidemiology 8:273–281.

Openshaw, S., A. W. Craft, M. Charlton, and J. M. Birch. 1988. Investigation of leukaemia clusters by use of a geographical analysis machine. Lancet 1:272–273.

Pike, M. C., and P. G. Smith. 1968. Disease clustering: A generalization of Knox's approach to the detection of space-time interactions. Biometrics 24:541–556.

Pike, M. C., and P. G. Smith. 1974. A case-control approach to examine diseases for evidence of contagion, including diseases with long latent periods. Biometrics 30:263–279.

Pinkel, D., and D. Nefzger. 1959. Some epidemiologic features of childhood leukemia in the Buffalo, New York area. Cancer 12:351–357.

Pinkel, D., J. E. Dowd, and I. D. J. Bross. 1963. Some epidemiological features of malignant solid tumors of children in Buffalo, N.Y. area. Cancer 16:28–33.

Raubertas, R. F. 1988. Spatial and temporal analysis of disease occurrence for detection of clustering. Biometrics 44:1121–1129.

Richardson, S., I. Stucker, and D. Hemon. 1987. Comparison of relative risks obtained in ecological and individual studies: Some methodological considerations. International Journal of Epidemiology 16:111–120.

Rothman, K.J. 1990a. A sobering start for the Cluster Buster's Conference. American Journal of Epidemiology 132(Suppl.):6–13.

Rothman, K. J. 1990b. No adjustments are needed for multiple comparisons. Epidemiology 1:43–46.

Rothwell, C. J., C. B. Hamilton, and P. E. Leaverton. 1991. Identification of sentinel health events as indicators of environmental contamination. Environmental Health Perspectives 94:261–263.

Rushton, G., D. Krishnamurti, R. Krishnamurty, and H. Song. 1995. A geographic information system analysis of urban infant mortality rates. Geographical Information Systems 5:52–56.

Rushton, G., R. Krishnamurty, D. Krishnamurti, P. Lolonis, and H. Song. 1996. The spatial relationship between infant mortality and birth defect rates in a U.S. city. Statistics in Medicine 15:1907–1919.

Sahu, S. K., R. B. Bendel, and C. P. Sison. 1993. Effect of relative risk and cluster configuration on the power of the one-dimensional scan statistic. Statistics in Medicine 12:1853–1865.

Savitz, D. A., and A. F. Olshan. 1995. Multiple comparisons and related issues in the interpretation of epidemiologic data. American Journal of Epidemiology 142:904–908.

Schulman, J., S. Selvin, and D. W. Merrill. 1988. Density equalized map projections: A method for analysing clustering around a fixed point. Statistics in Medicine 7:491–505.

Sexton, K., K. Olden, and B. L. Johnson. 1993. Environmental justice: The central role of research in establishing a credible scientific foundation for informed decision making. Toxicology and Industrial Health 9:685–727.

Shaw, G. M., S. Selvin, S. H. Swan, D. W. Merrill, and J. Schulman. 1988. An examination of three spatial disease clustering methodologies. International Journal of Epidemiology 17:913–919.

Snow, J. 1965. Snow on Cholera. New York: Hafner.

Stone, R. A. 1988. Investigations of excess environmental risks around putative sources: Statistical problems and a proposed test. Statistics in Medicine 7:649–660.

Susser, M. 1994a. The logic in ecological studies: I. The logic of analysis. American Journal of Public Health 84:825–829.

Susser, M. 1994b. The logic in ecological studies: II. The logic of design. American Journal of Public Health 84:830–835.

Symons, M. J., R. C. Grimson, and Y. C. Yuan. 1983. Clustering of rare events. Biometrics 39:193–205.

Tango, T. 1984. The detection of disease clusters in time. Biometrics 40:15–26.

Thomas, D., J. Siemiatycki, R. Dewar, J. Robins, M. Goldberg, and B. Armstrong. 1985. The problem of multiple inference in studies designed to generate hypotheses. American Journal of Epidemiology 122:1080–1095.

Turnbull, B. W., E. J. Iwano, W. J. Burnett, H. L. Howe, and L. C. Clark. 1990. Monitoring for clusters of disease: Application to leukemia incidence in upstate New York. American Journal of Epidemiology 132:S14–S22.

Wallenstein, S. 1980. A test for detection of clustering over time. American Journal of Epidemiology 104:576–584.

Wallenstein, S., J. Naus, and J. Glaz. 1993. Power of the scan statistic for detection of clustering. Statistics in Medicine 12:1829–1943.

Waller, L. A., and G. M. Jacquez. 1995. Disease models implicit in statistical tests of disease clustering. Epidemiology 6:584–590.

Waller, L. A., and A. B. Lawson. 1995. The power of focused tests to detect disease clustering. Statistics in Medicine 14:2291–2308.

Waller, L. A., B. W. Turnbull, L. C. Clark, and P. Nasca. 1994. Spatial pattern analyses to detect rare disease clusters. *In*: Case Studies in Biometry. N. Lange and L. Ryan, eds. New York: Wiley.

Walter, S. D. 1992a. The analysis of regional patterns in health data. I. Distributional considerations. American Journal of Epidemiology 136:730–741.

Walter, S. D. 1992b. The analysis of regional patterns in health data. II. The power to detect environmental effects. American Journal of Epidemiology 136:742–759.

Wartenberg, D. 1992. Screening for lead exposure using a geographic information system. Environmental Research, 59:310–317.

Wartenberg, D. 1994a. When Is a Cluster Really a Cluster? The Competing Agendas of Science, Society and Social Programs. Environmental Epidemiology: Science for Society or Science in Society? Hamilton, Ontario, Canada: McMaster University.

Wartenberg, D. 1994b. Use of geographic information systems for risk screening and epidemiology. *In*: J. S. Andrews, Jr., H. Frumkin, B. L. Johnson, M. A. Mehlman, C. Xintaras, and J. A. Bucsela, eds. Princeton, NJ: Princeton Scientific Publishing Co.

Wartenberg, D., and M. Greenberg. 1992. Methodological problems in investigating disease clusters. The Science of the Total Environment 127:173–185.

Wartenberg, D., and M. Greenberg. 1990b. Spatial models for detecting clusters of disease. *In*: Spatial Epidemiology. R. Thomas, ed. London: Pion.

Wartenberg, D., and M. Greenberg. 1993. Solving the cluster puzzle: Clues to follow and pitfalls to avoid. Statistics in Medicine 12:1763–1770.

Wartenberg, D., M. Greenberg, and R. Lathrop. 1993. Identification and characterization of populations living near high voltage transmission lines: A pilot study. Environmental Health Perspectives, 101:626–631.

Wartenberg, D., H. M. Kipen, P. F. Scully, and M. Greenberg. 1992. Racial oversight in occupational cancer epidemiology: A review of published studies. *In*: Johnson, B. L., Williams R. C., Harris C. M., eds. National Minority Health Conference: Focus on Environmental Contamination, pp. 137–147. Princeton, NJ: Princeton Scientific Publishing Co., Inc.

Weinstock, M. A. 1981. A generalized scan statistic test for the detection of clusters. International Journal of Epidemiology 10:289–293.

Whittemore, A. S., N. Friend, B. W. Brown, and E. A. Holly. 1987. A test to detect clusters of disease. Biometrika 74:631–635.

Whorton, D., R. Krauss, S. Marshal, and T. Milby. 1977. Infertility in male pesticide workers. Lancet 2:1259–1261.

Williams, G. W. 1984. Time-space clustering of disease. *In:* Statistical Methods for Cancer Studies. R. G. Cornell, ed. New York: Dekker.

Zimmerman, R. 1993. Social equity and environmental risk. Risk Analysis 13:649–666.

B

Acknowledgments

The committee would like to thank all those who took the time to attend and participate in its meetings and share their views with the committee either verbally or through written comments. All of those who participated in the committee's site visits and meetings are listed below.

Herbert K. Abrams
Department of Family and Community
  Medicine
University of Arizona Health Science
  Center

Anna Acuna
Living Is for Everyone

Rose Augustine
Tucsonians for a Clean Environment

Brett Baden
Harris Graduate School of Public
  Policy Studies
University of Chicago

Paul Baker
Pesticides Information Training
  Program
University of Arizona

Gerald S. Berenson
Tulane Center for Cardiovascular
  Health

John Bolcer
Hanford Health Information Archives
Foley Center Library
Gonzaga University

Dan S. Born
Louisiana Chemical Association

Robert Bostwick
Coeur D'Alene Tribe

Chuck Broscious
Environmental Defense Institute

Bunyan Bryant, Jr.
School of Natural Resources and
  Environment
University of Michigan

John Calcagni
Hanford Environmental Health
    Foundation

Cecilia Campillo
El Pueblo Clinic

James K. Carpenter
Office of Federal Programs

Luz Claudio
Community Outreach and
    Education Program
Mount Sinai Environmental Health
    Sciences Center

David Conrad
Department of Environmental
    Restoration and Waste
    Management
Nez Perce Tribe

Don Coursey
Harris School-University of Chicago

Kim Cunningham
Department of Environmental
    Restoration and Waste
    Management
Nez Perce Tribe

Margaret Wells Diaz
Environmental and Occupational
    Physician

Dianne Dugas
Louisiana Office of Public Health

Tim Flood
Office of Chronic Disease
    Epidemiology
Arizona Department of Health
    Services

Willie Fontenot
Public Protection Division
Louisiana Attorney General's Office

Jill Gay
Rural Health Coalition/Coalicion
    Rural

Deirdre Grace
Consortium for Risk Evaluation with
    Stakeholder Participants
School of Public Health and
    Community Medicine
University of Washington

Michael W. Grainey
Oregon Department of Energy

Marty Halper
Office of Environmental Justice
Environmental Protection Agency

Sharon Harrington
Mayor's Office of Environmental
    Affairs, New Orleans

Steven M. Hessl
Division of Occupational Medicine
Cook County Hospital

Carl Hild
Subsistence Division
Rural Alaska Community Action
    Program

Martha Holliday
Hanford Tribal Service Program
Indian Health Board

Daniel Hryhorczuk
Great Lakes Center for Occupational
    and Environmental Safety and
    Health
School of Public Health
University of Illinois

Barry L. Johnson
Centers for Disease Control and
  Prevention

Hazel Johnson
People for Community Recovery
  Environmental Organization

Karla Johnson
Environmental Protection Agency

Joyce C. Kelly
Environmental Protection Agency

Kaye H. Kilburn
University of Southern California
  School of Medicine

Deborah King
University of Akron

Lawrence P. King
Environmental Services
Summit County, Ohio

Robert Knox
Office of Environmental Justice
Environmental Protection Agency

Steve Kroll-Smith
Department of Sociology
Environmental Social Sciences
  Research Institute
University of New Orleans

Michael Lebowitz
University of Arizona Health Sciences
  Center

Edward B. Liebow
Battelle Seattle Research Center

John R. Lumpkin
Illinois Department of Public Health

Carlos Marentes
Sin Fronteras

Donna McDaniel
Deep South Center for Environmental
  Justice

John A. McLachlan
Tulane/Xavier Center for
  Bioenvironmental Research

John Middaugh
State of Alaska Department of Health
  and Social Services

Howard Mielke
College of Pharmacy
Xavier University

Marion Moses
Pesticide Education Center

Linda Ray Murray
Winfield Moody Health Center

Kenneth Olden
National Institute of Environmental
  Health Sciences

Gilbert S. Omenn
School of Public Health and
  Community Medicine
University of Washington

Mary Kay O'Rourke
Arizona Prevention Center
University of Arizona Health Sciences
  Center

Luis Ortega
Arizona Department of Health
  Services

Anthony Palazzo
The Pediatric Clinic

Nancy Parker
Hanford Health Information Network

William Pease
Environmental Defense Fund

Rafael Ponce
Hazards Consortium for Risk
    Evaluation with Stakeholder
    Participants
Department of Environmental Health
School of Public Health and
    Community Medicine
University of Washington

Donna Powaukee
Department of Environmental
    Restoration and Waste
    Management
Nez Perce Tribe

Max Power
Office of Nuclear and Mixed Waste
Washington State Department of
    Ecology

Dianne Quigley
George Perkins Marsh Institute
Clark University

James E. Rasmussen
Environmental Assurance Permits and
    Policy
National Environmental Policy Act

Florence Robinson
Southern University, Baton Rouge

Press Robinson
Southern University, Baton Rouge

Mark Robson
Environmental and Occupational
    Health Sciences Institute
Rutgers University

L. B. Sandy Rock
Hanford Health Information Network

Carol Roos
Sunnyside Unified School District 12
    Prevention Program

Jim Russell
Environmental Resource-Waste
    Management Program
Yakima Tribe

Virginia Sanchez
Citizen Alert, Native American
    Programs

Barbara Sattler
Environmental Health Education
    Center

Leslie Spino
Department of Natural Resources
Special Sciences and Resources

Marian Squeoch
Area Agency on Aging
Yakima Tribe

June Strickland
Psychosocial and Community Health
University of Washington School of
    Nursing

Timothy Takaro
Occupational and Environmental
    Medicine Clinic
University of Washington

Paul Templet
Institute for Environmental Studies
Louisiana State University

Sarah Moody Thomas
Stanley Scott Cancer Center
Louisiana State University Medical
  College

Emma Torres
Arizona Border Health Office

Robert G. Varady
Udall Center for Studies in Public
  Policy

Reuben C. Warren
Office of Minority Health
Centers for Disease Control and
  Prevention

Daniel Wartenberg
University of Medicine and Dentistry
  of New Jersey

Robert O. Washington
College of Urban and Public Affairs
University of New Orleans

Stephen West
Bureau of Environmental Health and
  Safety
Idaho Division of Health

Patricia Williams
Occupational Toxicology Outreach

Cynthia Williams-Mendy
Deep South Center for Environmental
  Justice

Robert Wolfe
Division of Subsistence
Alaska Department of Fish and Game

Beverly Wright
Deep South Center for Environmental
  Justice

Jill Zapian
Community Technical Assistance
Rural Health Office
Family and Community Medicine
Tucson, Arizona

C

Acronyms

ACS	American Chemical Society
COEP	Community Outreach and Education Program
EPA	Environmental Protection Agency
EMM	Ederer-Myers-Mantel
EOM	environmental and occupational medicine
GIS	geographic information systems
IOM	Institute of Medicine
LIFE	Living Is for Everyone
MCD	minor civil division
NIEHS	National Institute of Environmental Health Sciences
NRC	National Research Council
TCE	trichloroethylene
TEHIP	Toxicology and Environmental Health Information Program
TRI	Toxic Chemical Release Inventory
TSDF	treatment, storage, and disposal facility

D

Executive Order 12898:
Federal Actions to Address Environmental Justice in Minority Populations and Low-Income Populations

February 11, 1994

By the authority vested in me as President by the Constitution and the laws of the United States of America, it is hereby ordered as follows:

Sec. 1-1. Implementation

1-101. Agency Responsibilities

To the greatest extent practicable and permitted by law, and consistent with the principles set forth in the report on the National Performance Review, each Federal agency shall make achieving environmental justice part of its mission by identifying and addressing, as appropriate, disproportionately high and adverse human health or environmental effects of its programs, policies, and activities on minority populations and low-income populations in the United States and its territories and possessions, the District of Columbia, the Commonwealth of Puerto Rico, and the Commonwealth of the Mariana Islands.

1-102. Creation of an Interagency Working Group on Environmental Justice

a. Within 3 months of the date of this order, the Administrator of the Environmental Protection Agency ("Administrator") or the Administrator's designee shall convene an interagency Federal Working Group on Environmental Justice ("Working Group"). The Working Group shall comprise the heads of the following executive agencies and offices, or their designees:

    i. Department of Defense;
   ii. Department of Health and Human Services;
  iii. Department of Housing and Urban Development;
  iv. Department of Labor;
   v. Department of Agriculture;
  vi. Department of Transportation;
 vii. Department of Justice;
viii. Department of the Interior;
  ix. Department of Commerce;
   x. Department of Energy;
  xi. Environmental Protection Agency;
 xii. Office of Management and Budget;
xiii. Office of Science and Technology Policy;
xiv. Office of the Deputy Assistant to the President for Environmental Policy;
 xv. Office of the Assistant to the President for Domestic Policy;
xvi. National Economic Council;
xvii. Council of Economic Advisers; and
xviii. such other Government officials as the President may designate.

The Working Group shall report to the President through the Deputy Assistant to the President for Environmental Policy and the Assistant to the President for Domestic Policy.

b. The Working Group shall:

    1. provide guidance to Federal agencies on criteria for identifying disproportionately high and adverse human health or environmental effects on minority populations and low-income populations;
    2. coordinate with, provide guidance to, and serve as a clearinghouse for, each Federal agency as it develops an environmental justice strategy as required by section 1-103 of this order, in order to ensure that the administration, interpretation, and enforcement of programs, activities, and policies are undertaken in a consistent manner;
    3. assist in coordinating research by, and stimulating cooperation among, the Environmental Protection Agency, the Department of Health and Human Services, the Department of Housing and Urban Development, and other agencies conducting research or other activities in accordance with section 3-3 of this order;
    4. assist in coordinating data collection, required by this order;
    5. examine existing data and studies on environmental justice;
    6. hold public meetings as required in section 5-502(d) of this order; and
    7. develop interagency model projects on environmental justice that evidence cooperation among Federal agencies.

1-103. Development of Agency Strategies

a. Except as provided in section 6-605 of this order, each Federal agency shall develop an agency-wide environmental justice strategy, as set forth in subsections (b)-(e) of this section that identifies and addresses disproportionately high and adverse human health or environmental effects of its programs, policies, and activities on minority populations and low-income populations. The environmental justice strategy shall list programs, policies, planning and public participation processes, enforcement, and/or rulemakings related to human health or the environment that should be revised to, at a minimum:

    1. promote enforcement of all health and environmental statutes in areas with minority populations and low-income populations;
    2. ensure greater public participation;
    3. improve research and data collection relating to the health of and environment of minority populations and low-income populations; and
    4. identify differential patterns of consumption of natural resources among minority populations and low-income populations.

In addition, the environmental justice strategy shall include, where appropriate, a timetable for undertaking identified revisions and consideration of economic and social implications of the revisions.

b. Within 4 months of the date of this order, each Federal agency shall identify an internal administrative process for developing its environmental justice strategy, and shall inform the Working Group of the process.
c. Within 6 months of the date of this order, each Federal agency shall provide the Working Group with an outline of its proposed environmental justice strategy.
d. Within 10 months of the date of this order, each Federal agency shall provide the Working Group with its proposed environmental justice strategy.
e. Within 12 months of the date of this order, each Federal agency shall finalize its environmental justice strategy and provide a copy and written description of its strategy to the Working Group. During the 12–month period from the date of this order, each Federal agency, as part of its environmental justice strategy, shall identify several specific projects that can be promptly undertaken to address particular concerns identified during the development of the proposed environmental justice strategy, and a schedule for implementing those projects.
f. Within 24 months of the date of this order, each Federal agency shall report to the Working Group on its progress in implementing its agency-wide environmental justice strategy.
g. Federal agencies shall provide additional periodic reports to the Working Group as requested by the Working Group.

1-104. Reports to the President

Within 14 months of the date of this order, the Working Group shall submit to the President, through the Office of the Deputy Assistant to the President for Environmental Policy and the Office of the Assistant to the President for Domestic Policy, a report that describes the implementation of this order, and includes the final environmental justice strategies described in section 1-103(e) of this order.

Sec. 2-2. Federal Agency Responsibilities for Federal Programs

Each Federal agency shall conduct its programs, policies, and activities that substantially affect human health or the environment, in a manner that ensures that such programs, policies, and activities do not have the effect of excluding persons (including populations) from participation in, denying persons (including populations) the benefits of, or subjecting persons (including populations) to discrimination under, such programs, policies, and activities, because of their race, color, or national origin.

Sec. 3-3. Research, Data Collection, and Analysis

3-301. Human Health and Environmental Research and Analysis

a. Environmental human health research, whenever practicable and appropriate, shall include diverse segments of the population in epidemiological and clinical studies, including segments at high risk from environmental hazards, such as minority populations, low-income populations, and workers who may be exposed to substantial environmental hazards.
b. Environmental human health analyses, whenever practicable and appropriate, shall identify multiple and cumulative exposures.
c. Federal agencies shall provide minority populations and low-income populations the opportunity to comment on the development and design of research strategies undertaken pursuant to this order.

3-302. Human Health and Environmental Data Collection and Analysis

To the extent permitted by existing law, including the Privacy Act, as amended (5 U.S.C. section 552a):

a. Each Federal agency, whenever practicable and appropriate, shall collect, maintain, and analyze information assessing and comparing environmental and human health risks borne by populations identified by race, national origin, or income. To the extent practical and appropriate, Federal agencies shall use this information to determine whether their programs, policies, and activities have disproportionately high and adverse human health or environmental effects on minority populations and low-income populations;

b. In connection with the development and implementation of agency strategies in section 1-103 of this order, each Federal agency, whenever practicable and appropriate, shall collect, maintain, and analyze information on the race, national origin, income level, and other readily accessible and appropriate information for areas surrounding facilities or sites expected to have a substantial environmental, human health, or economic effect on the surrounding populations, when such facilities or sites become the subject of a substantial Federal environmental administrative or judicial action. Such information shall be made available to the public, unless prohibited by law; and

c. Each Federal agency, whenever practicable and appropriate, shall collect, maintain, and analyze information on the race, national origin, income level, and other readily accessible and appropriate information for areas surrounding Federal facilities that are:

1. subject to the reporting requirements under the Emergency Planning and Community Right-to-Know Act, 42 U.S.C. section 11001-11050 as mandated in Executive Order No. 12856; and

2. expected to have a substantial environmental, human health, or economic effect on surrounding populations. Such information shall be made available to the public, unless prohibited by law.

d. In carrying out the responsibilities in this section, each Federal agency, whenever practicable and appropriate, shall share information and eliminate unnecessary duplication of efforts through the use of existing data systems and cooperative agreements among Federal agencies and with State, local, and tribal governments.

Sec. 4-4. Subsistence Consumption of Fish and Wildlife

4-401. Consumption Patterns

In order to assist in identifying the need for ensuring protection of populations with differential patterns of subsistence consumption of fish and wildlife, Federal agencies, whenever practicable and appropriate, shall collect, maintain, and analyze information on the consumption patterns of populations who principally rely on fish and/or wildlife for subsistence. Federal agencies shall communicate to the public the risks of those consumption patterns.

4-402. Guidance

Federal agencies, whenever practicable and appropriate, shall work in a coordinated manner to publish guidance reflecting the latest scientific information available concerning methods for evaluating the human health risks associated with the consumption of pollutant-bearing fish or wildlife. Agencies shall consider such guidance in developing their policies and rules.

Sec. 5-5. Public Participation and Access to Information

a.   The public may submit recommendations to Federal agencies relating to the incorporation of environmental justice principles into Federal agency programs or policies. Each Federal agency shall convey such recommendations to the Working Group.
b.   Each Federal agency may, whenever practicable and appropriate, translate crucial public documents, notices, and hearings relating to human health or the environment for limited English–speaking populations.
c.   Each Federal agency shall work to ensure that public documents, notices, and hearings relating to human health or the environment are concise, understandable, and readily accessible to the public.
d.   The Working Group shall hold public meetings, as appropriate, for the purpose of fact-finding, receiving public comments, and conducting inquiries concerning environmental justice. The Working Group shall prepare for public review a summary of the comments and recommendations discussed at the public meetings.

Sec. 6-6. General Provisions

6-601. Responsibility for Agency Implementation

The head of each Federal agency shall be responsible for ensuring compliance with this order. Each Federal agency shall conduct internal reviews and take such other steps as may be necessary to monitor compliance with this order.

6-602. Executive Order No. 12250

This Executive order is intended to supplement but not supersede Executive Order No. 12250, which requires consistent and effective implementation of various laws prohibiting discriminatory practices in programs receiving Federal financial assistance. Nothing herein shall limit the effect or mandate of Executive Order No. 12250.

6-603. Executive Order No. 12875

This Executive order is not intended to limit the effect or mandate of Executive Order No. 12875.

6-604. Scope

For purposes of this order, Federal agency means any agency on the Working Group, and such other agencies as may be designated by the President, that conduct any Federal program or activity that substantially affects human health or the environment. Independent agencies are requested to comply with the provisions of this order.

6-605. Petitions for Exemptions

The head of a Federal agency may petition the President for an exemption from the requirements of this order on the grounds that all or some of the petitioning agency's programs or activities should not be subject to the requirements of this order.

6-606. Native American Programs

Each Federal agency responsibility set forth under this order shall apply equally to Native American programs. In addition, the Department of the Interior, in coordination with the Working Group, and, after consultation with tribal leaders, shall coordinate steps to be taken pursuant to this order that address Federally recognized Indian Tribes.

6-607. Costs

Unless otherwise provided by law, Federal agencies shall assume the financial costs of complying with this order.

6-608. General

Federal agencies shall implement this order consistent with, and to the extent permitted by, existing law.

6-609. Judicial Review

This order is intended only to improve the internal management of the executive branch and is not intended to, nor does it create any right, benefit, or trust responsibility, substantive or procedural, enforceable at law or equity by a party against the United States, its agencies, its officers, or any person. This order shall not be construed to create any right to judicial review involving the compliance or noncompliance of the United States, its agencies, its officers, or any other person with this order.

William Clinton
The White House,
February 11, 1994.

Exec. Order No. 12898
59 FR 7629
1994 WL 43891 (Pres.)

THE WHITE HOUSE

WASHINGTON

February 11, 1994

MEMORANDUM FOR THE HEADS OF ALL DEPARTMENTS
AND AGENCIES

SUBJECT: Executive Order on Federal Actions to Address
Environmental Justice in Minority Populations and Low-Income Populations

Today I have issued an Executive Order on Federal Actions to Address Environmental Justice in Minority Populations and Low-Income Populations. That order is designed to focus Federal attention on the environmental and human health conditions in minority communities and low-income communities with the goal of achieving environmental justice. That order is also intended to promote nondiscrimination in Federal programs substantially affecting human health and the environment, and to provide minority communities and low-income communities access to public information on, and an opportunity for public participation in, matters relating to human health or the environment.

The purpose of this separate memorandum is to underscore certain provisions of existing law that can help ensure that all communities and persons across this Nation live in a safe and healthful environment. Environmental and civil rights statutes provide many opportunities to address environmental hazards in minority communities and low-income communities. Application of these existing statutory provisions is an important part of this Administration's efforts to prevent those minority communities and low-income communities from being subject to disproportionately high and adverse environmental effects.

I am therefore today directing that all department and agency heads take appropriate and necessary steps to ensure that the following specific directives are implemented immediately:

In accordance with Title VI of the Civil Rights Act of 1964, each Federal agency shall ensure that all programs or activities receiving Federal financial assistance that affect human health or the environment do not directly, or through contractual or other arrangements, use criteria, methods, or practices that discriminate on the basis of race, color, or national origin.

Each Federal agency shall analyze the environmental effects, including human health, economic, and social effects, of Federal actions, including effects on minority communities and low-income communities, when such analysis is required by the National Environmental Policy Act of 1969 (NEPA), 42 U.S.C.

section 4321 et seq. Mitigation measures outlined or analyzed in an environmental assessment, environmental impact statement, or record of decision, whenever feasible, should address significant and adverse environmental effects of proposed Federal actions on minority communities and low-income communities.

Each Federal agency shall provide opportunities for community input in the NEPA process, including identifying potential effects and mitigation measures in consultation with affected communities and improving the accessibility of meetings, crucial documents, and notices.

The Environmental Protection Agency, when reviewing environmental effects of proposed action of other Federal agencies under section 309 of the Clean Air Act, 42 U.S.C. section 7609, shall ensure that the involved agency has fully analyzed environmental effects on minority communities and low-income communities, including human health, social, and economic effects.

Each Federal agency shall ensure that the public, including minority communities and low-income communities, has adequate access to public information relating to human health or environmental planning, regulations, and enforcement when required under the Freedom of Information Act, 5 U.S.C. section 552, the Sunshine Act, 5 U.S.C. section 552b, and the Emergency Planning and Community Right-to-Know Act, 42 U.S.C. section 11044.

This memorandum is intended only to improve the internal management of the Executive Branch and is not intended to, nor does it create, any right, benefit, or trust responsibility, substantive or procedural, enforceable at law or equity by a party against the United States, its agencies, its officers, or any person.

                                                                    William Clinton

E

Committee and Staff Biographies

COMMITTEE BIOGRAPHIES

JAMES R. GAVIN III (*Cochair*) is a Senior Scientific Officer of the Howard Hughes Medical Institute in Chevy Chase, Maryland. Prior to joining the senior scientific staff at the Hughes Institute in October 1991, he served as William K. Warren Professor for Diabetes Studies; Professor of Medicine; and Chief, Diabetes Section at the University of Oklahoma Health Sciences Center. He completed his B.S. in chemistry at Livingstone College, a Ph.D. in biochemistry at Emory University, and his M.D. at Duke University Medical School. He did postgraduate medical training in internal medicine and endocrinology at Barnes Hospital in the Washington University Medical Center, St. Louis, Missouri. He was Associate Professor of Medicine at Washington University prior to his move to Oklahoma in January 1987. He is a member of the Endocrine Society, the American Diabetes Association, the American Society for Clinical Investigation, the American Association of Physicians, and the American Association of Academic Minority Physicians. He serves on the editorial board of *Academic Medicine*, is Past President of the American Diabetes Association, and serves as a member of the National Institute of Diabetes and Digestive and Kidney Diseases Advisory Council. He has received numerous civic and academic awards and honors, including being voted Clinical Teacher of the Year at Barnes Hospital, Outstanding Clinician in the field of Diabetes by the American Diabetes Association, and Internist of the Year by the National Medical Association. He is a recipient of the 1998 Emory University Medal for Distinguished Achievement. He has served on a broad range of study sections, national boards, and advisory groups in the private and governmental sectors. He presently serves as Senior Program Consultant and Director of the Minority Medical Faculty Development Program of the Robert Wood Johnson Foundation (RWJF) and is also a member of the Board of Trustees of RWJF. His research interests have been in insulin resistance and diabetes mellitus. He has authored more than 160 original papers, book chapters, and scientific abstracts. Dr. Gavin is a member of the Institute of Medicine.

DONALD R. MATTISON (*Cochair*) is Medical Director of the March of Dimes Birth Defects Foundation. From 1990 to 1998 he was Dean of the Graduate School of Public Health and Professor of Environmental and Occupational Health and Obstetrics and Gynecology at the University of Pittsburgh. Dr. Mattison received his M.D. from The College of Physicians and Surgeons, Columbia University, and clinical training in obstetrics and gynecology at Sloane Hospital for Women, Columbia Presbyterian Medical Center, in New York City. Dr. Mattison obtained postgraduate research training at the National Institutes of Health. From 1984 to 1990 he was Professor of Obstetrics and Gynecology and Toxicology at the University of Arkansas for Medical Sciences. During this period he was Acting Director of the Human Risk Assessment Program at the National Center for Toxicological Research, a component of the U.S. Food and Drug Administration. Dr. Mattison is a member of many local and national boards. He has published more than 140 papers, chapters, and reviews in the areas of reproductive and developmental toxicology, risk assessment, and clinical obstetrics and gynecology.

REGINA AUSTIN is William A. Schnader Professor of Law at the University of Pennsylvania. She received a B.A. from the University of Rochester in 1970 and a J.D. from the University of Pennsylvania in 1973. She is a member of the Order of the Coif, the legal honorary society. Before joining the University of Pennsylvania faculty in 1977, Professor Austin was an associate with the firm of Schnader, Harrison, Segal & Lewis. She has been a visiting professor at Harvard and Stanford law schools. Professor Austin has written on various topics including the working conditions of low-status minority and female workers, the construction of black economic activity as deviance, and the minority grassroots environmental movement.

DAVID R. BAINES is a practicing family physician on the Coeur d'Alene Indian Reservation in rural Idaho. He received his medical degree from Mayo Medical School in Rochester, Minnesota. Since 1992, Dr. Baines has served as a member of the Commission on Membership and Members Services of the American Academy of Family Physicians (AAFP). In 1990 he chaired the AAFP Committee on Minority Health Affairs and has been a consultant to the AAFP Subcommittee on Indian Health. Since 1984 he has served as the liaison between AAFP and the Association of American Indian Physicians. He was President of the Association of American Indian Physicians in 1990 and 1991. He has served as a consultant to National Institutes of Health committees, a member of Institute of Medicine committees, and a member of the National Academy of Sciences Polar Research Board. In 1995 he was selected by Donna Shalala, Secretary of Health and Human Services, to serve on the Clinical Laboratory Improvement Advisory Committee.

BARUCH FISCHHOFF is University Professor in the Department of Engineering and Public Policy and the Department of Social and Decision Sciences, Carnegie Mellon University. He holds a B.S. in mathematics from Wayne State University and an M.A. and Ph.D. in psychology from the Hebrew University of Jerusalem. He is a fellow of the American Psychological Association and recipient of its Early Career Awards for Distinguished Scientific Contribution to Psychology (1980) and for Contributions to Psychology in the Public Interest (1991). He is a fellow of the Society for Risk Analysis, as well as recipient of its Distinguished Achievement Award (1991). He has served on many Institute of Medicine and National Research Council panels, including the Committee on the NIH Priority-Setting Process, the Commission on Behavioral and Social Sciences and Education, and the Environmental Health Sciences Roundtable. He serves on several editorial boards, including the Journal of Risk and Uncertainty and Journal of Experimental Psychology: Applied. Dr. Fischhoff is a member of the Institute of Medicine.

GEORGE FRIEDMAN-JIMÉNEZ is founder and Director of the Occupational and Environmental Medicine Clinic at Bellevue Hospital of the New York University School of Medicine, and epidemiologist in the Center for Urban Epidemiologic Studies of the New York Academy of Medicine. Dr. Friedman-Jiménez received his medical degree from the Albert Einstein College of Medicine. He is board certified in internal medicine and preventive medicine (occupational) and has completed a postdoctoral fellowship in epidemiology at the Columbia University School of Public Health. His research interests include the occupational and environmental epidemiology of asthma and methods of clinical epidemiology. Dr. Friedman-Jiménez is the recipient of an Academic Award in Environmental and Occupational Medicine from the National Institute of Environmental Health Sciences (NIEHS) and a Visiting Scholar in Residence research fellowship from the National Institute for Occupational Safety and Health and Morehouse School of Medicine. He is a member of the advisory board of the National Hispanic Medical Association and serves on the Board of Scientific Counselors of the National Toxicology Program of NIEHS, as well as on the Biennial Report on Carcinogens Subcommittee.

BERNARD D. GOLDSTEIN is Director of the Environmental and Occupational Health Sciences Institute, a joint program of Rutgers, The State University of New Jersey, and the University of Medicine and Dentistry of New Jersey (UMDNJ) Robert Wood Johnson Medical School. He is Chair of the Department of Environmental and Community Health, UMDNJ-Robert Wood Johnson Medical School. He received his B.S. from the University of Wisconsin and his M.D. from New York University School of Medicine. Dr. Goldstein is board certified in internal medicine and hematology and board certified in toxicology. Dr. Goldstein served as Assistant Administrator for Research and Development, Environmental Protection Agency (EPA) from 1983 to 1985, was a member and Chair of the NIH Toxicology Study Section, and was on EPA's Clean Air Scientific Advisory Committee. He has published more than 200 articles and book chapters related to environmental health sciences and public policy. Dr. Goldstein is a member of the Institute of Medicine.

JAMES G. HAUGHTON is Medical Director of Public Health Programs and Services in the Los Angeles County Department of Health Services. He received his M.D. from Loma Linda University College of Medicine and his M.P.H. from the Columbia University School of Public Health. His professional activities have included the practice of obstetrics and gynecology and, for the past 35 years, the leadership and management of a large public hospital and public health agencies. He has served as Executive Medical Director of the New York City Department of Public Health; Chief Executive Officer of the Health and Hospitals Governing Commission of Cook County, Illinois; Executive Vice President of the Charles Drew Postgraduate Medical School; Director of Public Health and Human Services for the city of Houston, Texas; and Medical Director and Chief-of-Staff of Los Angeles County's King/Drew Medical Center. Dr. Haughton is a member of the Institute of Medicine and the National Academy of Social Insurance. He received the Humanitarian Award from the National Association of Health Services Executives and an honorary Doctor of Sciences degree from the University of Health Sciences/Chicago Medical College. He serves on the Board of Directors of the California Conference of Local Health Officers. Dr. Haughton has written many articles regarding health care for the poor and medically indigent.

SANDRAL HULLETT is Executive Director of West Alabama Health Services, a community health center located in rural west Alabama. She has a bachelor's degree from Alabama A&M University in Normal; a medical degree from the Medical College of Pennsylvania, Philadelphia; and a master's in public health from the University of Alabama at Birmingham. Since completing a residency in family practice and fulfilling a National Health Services Corps obligation, Dr. Hullett developed an interest in rural health care, including health care planning and delivery to the underserved, underinsured, and poor of this area. Dr. Hullett is a member of the Board of Trustees of the University of Alabama System and has been appointed a member of the Practicing Physicians Advisory Council of the U.S. Department of Health and Human Services. Dr.

Hullett is a member of the Institute of Medicine and the Alabama Health Care Reform Task Force.

LOVELL A. JONES is Director of Experimental Gynecology-Endocrinology of the Department of Gynecologic Oncology at the University of Texas M. D. Anderson Cancer Center and has held that position since 1988. He is also a Professor of Gynecologic Oncology, Biochemistry and Molecular Biology at the same institution. Prior to joining the faculty at M. D. Anderson, he was a National Cancer Institute postdoctoral fellow and then a lecturer in the Department of Obstetrics, Gynecology and Reproductive Sciences at the University of California Medical Center-San Francisco. Dr. Jones holds a B.S. degree in biological sciences from California State University at Hayward and M.S. and Ph.D. degrees in zoology with an emphasis in tumor biology and endocrinology from the University of California at Berkeley. Dr. Jones has served on a number of national advisory committees including the National Advisory Council for Environmental Health Sciences at the National Institute of Environmental Health Sciences. He presently serves on one of the Environmental Protection Agency's Scientific Advisory Panels and the U.S. Department of Defense Breast Cancer Research Integration Panel. Dr. Jones has contributed to many publications and journals in the area of hormonal carcinogenesis. He is also a well-respected expert in the area of minorities and cancer. Dr. Jones is a founding co-chair of the Intercultural Cancer Council and the founder of the Biennial Symposium Series on Minorities, the Medically Underserved and Cancer.

CHARLES LEE is Director of Environmental Justice for the United Church of Christ Commission for Racial Justice. He has played a singularly pioneering role in the definition and development of issues of race and the environment. Mr. Lee is the architect of the two seminal national events in the emergence of environmental justice as a nationally prominent issue: the landmark 1987 report *Toxic Wastes and Race in the United States* and the historic 1991 National People of Color Environmental Leadership Summit. He has served on many advisory panels and boards, including the Environmental Protection Agency's National Environmental Justice Advisory Council, where he chaired the Waste and Facility Siting Subcommittee. He has written numerous articles, reports, and scholarly papers and is the editor of several books. Mr. Lee is a leading national advocate for empowered community involvement in environmental research and policy development.

ROGER O. McCLELLAN serves as President of the Chemical Industry Institute of Toxicology located in Research Triangle Park, North Carolina. He has held that position since 1988. He is well known for his work in the related fields of toxicology and risk assessment, especially concerning the potential human risks of airborne materials. Dr. McClellan has previously served as President of the Society of Toxicology and American Association for Aerosol Research. He is a fellow of the Society for Risk Analysis and an elected member of the Institute of Medicine of the National Academy of Sciences. He has served in an advisory role to many public and private organizations including Chair of the Environmental Protection Agency's Clean Air Scientific Advisory Committee and the National Research Council's Committee on Toxicology. Dr. McClellan is a strong advocate for the need to integrate data from epidemiologic, controlled clinical, laboratory animal, and cell studies to assess the human health risks of occupational or environmental exposure to chemicals.

MARY ANN SMITH has held the position of Assistant Professor of Environmental Sciences (Toxicology) at the University of Texas Health Science Center at Houston School of Public Health from 1991 to the present. She received a B.S. in pharmacology from the University of Texas at Austin in 1979 and received her Ph.D. in 1984. Before accepting her current position at the University of Texas, Dr. Smith was an Assistant Professor of Pharmacy (Pharmacology/ Toxicology) at the University of New Mexico, Albuquerque. Early in her career she was a staff fellow at the Laboratory of Experimental Therapeutics and Metabolism at the National Cancer Institute, National Institutes of Health, in Bethesda, Maryland. The SmithKline & French Laboratories, Department of Pharmacology/Toxicology, Philadelphia College of Pharmacy and Science in Philadelphia, Pennsylvania, offered Dr. Smith a postdoctoral fellowship from 1985 to 1987. Dr. Smith received the University of New Mexico Graduate and Undergraduate Teacher of the Year Award for 1989–1990.

WALTER J. WADLINGTON is James Madison Professor of Law at the University of Virginia School of Law and Professor of Legal Medicine at the University of Virginia School of Medicine. He was the Program Director of the Robert Wood Johnson Foundation Medical Malpractice Program and also served on the Clinical Scholars Advisory Board. He is a member of the Institute of Medicine and the American Law Institute, and a Trustee of the Educational Commission for Foreign Medical Graduates. He teaches in the areas of law and medicine, family law, and children's health care.

INSTITUTE OF MEDICINE STAFF

MARY JAY BALL-HENDERSON was a Project Assistant in the Institute of Medicine's Division of Health Sciences Policy. During her 5 years at the Institute of Medicine, she worked with several committees, including the Committee on Assessing Genetic Risks, the Forum on Drug Development, the Forum on Blood Safety and Blood Availability, the Committee on Strengthening the Geriatric Content of Medical Training, the Committee on Xenograft Transplantation: Ethical Issues and Public Policy, and the Committee on the Future of Academic Health Centers. Prior to working at the Institute of Medicine, she worked for 5 years at the National Academy of Engineering. She left the Institute of Medicine in January 1997.

YVETTE J. BENJAMIN was a Research Associate in the Institute of Medicine's Division of Health Sciences Policy. She received training as a physician's assistant at George Washington University and holds a B.A. in psychology from Incarnate Word College in San Antonio, Texas, and a B.S. in biology from George Washington University. In 1995 she received a master's degree in public health from George Washington University, with a concentration in health policy. Ms. Benjamin has had extensive experience both as a clinician and a researcher in the area of AIDS and human immunodeficiency virus. Ms. Benjamin was at the Institute of Medicine from 1993 to October 1998 and has provided support for several studies in the areas of xenograft transplantation and military nursing research and has conducted a workshop on the future of academic health centers.

PETER BOUXSEIN is a Senior Program Officer in the Institute of Medicine. Mr. Bouxsein has an undergraduate degree in science from Carnegie Mellon University and a law degree from the University of Chicago. He has 23 years of service with the federal government, including the Department of Justice, the Office of Economic Opportunity, the Health Care Financing Administration, and the Agency for Health Care Policy and Research, in the areas of civil rights, higher education, and health care. Seven of those years were spent as counsel to the Subcommittee on Health and the Environment, U.S. House of Representatives, focusing on Medicare, health care technology, graduate medical education, and clinical research. In addition, Mr. Bouxsein has served as the Deputy Director of the Institute for Public Policy Studies, University of Michigan, and Deputy Executive Vice President of the American College of Physicians. He is also a research associate and lecturer at the Johns Hopkins School of Public Health.

LINDA A. DEPUGH is Administrative Assistant for the Health Sciences Section of the Institute of Medicine. Ms. Depugh has more than 27 years of experience working in the Academy complex. She served as Administrative Assistant for the Division of Health Promotion and Disease Prevention for several years prior to joining the Division of Health Sciences Policy in 1994. Linda provided administrative assistance to the Board on Health Sciences Policy and the Division of Health Promotion and Disease Prevention by coordinating specific tasks that are crucial to the progress and completion of program activities. She obtained her associate's degree from Durham Business College in Durham, North Carolina.

CHARLES H. EVANS, JR., is Head of the Health Sciences Section in the Institute of Medicine. Dr. Evans joined the staff of the Institute of Medicine in March 1998. As Head of the new Health Sciences Section, Dr. Evans has management responsibility for all scientific, administrative, and financial affairs of the Health Sciences Section, which includes the Health Sciences Policy Program and the Neuroscience and Behavioral Health Program and their respective boards in the Institute of Medicine. Dr. Evans is a pediatrician and immunologist and holds the rank of Captain (retired) in the U.S. Public Health Service with 27 years of service as a medical scientist at the National Institutes of Health. He received his B.S. (biology) degree from Union College in 1962 and M.D. and Ph.D. (microbiology) degrees from the University of Virginia in 1969. He was an intern and resident in pediatrics at the University of Virginia from 1969 to 1971, and from 1971 to 1998 he served as a Medical Officer in the U.S. Public Health Service Commissioned Corps and, concurrently from 1976 to 1998, was Chief of the Tumor Biology Section at the National Cancer Institute. An expert in carcinogenesis and the normal immune system defenses to the development of cancer, he is the author of more than 250 scientific publications. He and his laboratory colleagues discovered the cytokine leukoregulin in 1983 and were awarded three U.S. patents. Dr. Evans has been active as an adviser to community medicine and higher education through his service on the Board of Trustees of Suburban Hospital Health System (1988 to present) and on the Arts and Sciences Alumni Council at the University of Virginia (1987 to 1997). He is the recipient of numerous scientific awards including the Outstanding Service Medal from the U.S. Public Health Service and the Wellcome Medal and Prize. Dr. Evans has been a member of the editorial boards of several scientific journals, has served on a variety of scientific advisory committees, and is a fellow of the American Institute of Chemists and a credentialed fellow in Health Systems Administration of the American Academy of Medical Administrators.

EDWARD HILL III was Senior Program Officer and Study Director in the Institute of Medicine's Division of Health Sciences Policy. Dr. Hill received his J.D. degree from the University of Baltimore School of Law and his M.D. degree from Meharry Medical College in Nashville, Tennessee. He has specialized training in medical law for the attending physician and has consulted for several health care and law firms. He left the Institute of Medicine in November 1996.

ANDREW M. POPE is Director of the Health Sciences Policy Program at the Institute of Medicine. With expertise in physiology, toxicology, and epidemiology, his primary interests focus on environmental and occupational influences on human health. As a research fellow in the Division of Pharmacology and Toxicology at the U.S. Food and Drug Administration, Dr. Pope's research focused on the biochemical, neuroendocrine, and reproductive effects of various environmental substances on food-producing animals. During his tenure at the

National Academy of Sciences and since 1989 at the Institute of Medicine, Dr. Pope has directed studies on and edited numerous reports on environmental and occupational issues; topics include injury control, disability prevention, biologic markers, neurotoxicology, indoor allergens, and the inclusion of environmental and occupational health content in medical and nursing school curricula.

VALERIE PETIT SETLOW was Director of the Division of Health Sciences Policy. Dr. Setlow received her B.S. in chemistry from Xavier University and her Ph.D. in molecular biology from Johns Hopkins University. Dr. Setlow has conducted research in molecular hematology and virology and has had a distinguished career in government, serving in numerous positions including Director of the Cystic Fibrosis Research Programs at the National Institutes of Health and, in her last position, as Acting Director of the National AIDS Program Office. Dr. Setlow left the Institute of Medicine in December 1997 to become Deputy Director of the Tulane/Xavier Center for Bioenvironmental Research.

GLEN SHAPIRO is a Project Assistant and Research Assistant in the Health Sciences Policy Program. He is also providing support for the Committee on Fluid Resuscitation for Combat Casualties and the Roundtable on Environmental Health Sciences, Research, and Medicine. Glen earned his bachelor's degree in Russian language and literature at Wesleyan University, Middletown, Connecticut.

JAMAINE L. TINKER was a Financial Associate with the Institute of Medicine of the National Academy of Sciences. She provided support over the course of research projects by performing financial and administrative oversight activities. She worked closely with program staff to prepare proposal budgets, cost projections, financial reports, and analyses. In this capacity, she often served as the liaison between the division and other academy offices such as Contracts and Grants, Accounting, Purchasing, Payroll, Travel Services, and Human Resources. Jamaine earned a business administration certificate in 1994 from Georgetown University and a liberal arts degree in 1987 from Wittenberg University, Springfield, Ohio, where she majored in Spanish and minored in mathematics.

Index